T0073503

About Island Press

Since 1984, the nonprofit organization Island Press has been stimulating, shaping, and communicating ideas that are essential for solving environmental problems worldwide. With more than 1,000 titles in print and some 30 new releases each year, we are the nation's leading publisher on environmental issues. We identify innovative thinkers and emerging trends in the environmental field. We work with world-renowned experts and authors to develop cross-disciplinary solutions to environmental challenges.

Island Press designs and executes educational campaigns in conjunction with our authors to communicate their critical messages in print, in person, and online using the latest technologies, innovative programs, and the media. Our goal is to reach targeted audiences—scientists, policymakers, environmental advocates, urban planners, the media, and concerned citizens—with information that can be used to create the framework for long-term ecological health and human well-being.

Island Press gratefully acknowledges major support of our work by The Agua Fund, The Andrew W. Mellon Foundation, The Bobolink Foundation, The Curtis and Edith Munson Foundation, Forrest C. and Frances H. Lattner Foundation, The JPB Foundation, The Kresge Foundation, The Oram Foundation, Inc., The Overbrook Foundation, The S.D. Bechtel, Jr. Foundation, The Summit Charitable Foundation, Inc., and many other generous supporters.

The opinions expressed in this book are those of the author(s) and do not necessarily reflect the views of our supporters.

Energy Democracy

Energy Democracy
Advancing Equity in Clean Energy Solutions

edited by
DENISE FAIRCHILD AND AL WEINRUB

Washington | Covelo | London

Library of Congress Control Number: 2017935090

All Island Press books are printed on environmentally responsible materials.

Manufactured in the United States of America
10 9 8 7 6 5 4 3 2 1

Keywords: Centralized renewable energy, clean energy, climate activism, climate change, climate justice, climate solutions, community development, community power, decentralized renewable energy, economic development, economic justice, ecosystem, energy, energy activism, energy commons, energy cooperatives, energy democracy, energy equity, energy sources, energy transition, environmental activism, environmental justice, just transition, labor unions, new economy, racial justice, regenerative economy, renewable energy, resilient communities, social equity, social justice, sustainable economy

Contents

CONTENTS

Prologue

We put this book together to shine a spotlight on efforts to democratize energy in the United States—to give voice to the organizations and individuals leading the way to a transformed energy future. We hope that by giving expression to their vision, strategies, organizing efforts, and development models we will inspire others to join the growing energy democracy movement.

We were propelled by the urgent necessity—the existential necessity—for human society to create a new kind of energy future. The growth and dominance of fossil fuel energy over the last 150 years has had profound adverse environmental, economic, and social impacts. As a result, our very survival hangs in the balance.

We wanted this book to look critically at how our use of energy has driven ecosystem destruction, economic insecurity, and social injustice, and, at the same time, we hoped that it would promote the new energy paradigm and decentralized energy model needed for a sustainable future.

This is the work of the energy democracy movement.

We see this work as of utmost importance at this critical moment in history. Driven by a changing climate, the transition from fossil fuels to renewable energy is creating uncertainty and panic for the energy establishment—the large corporate energy producers, utility monopolies, and federal and state government agencies that serve the status quo. At the same time, the global economic system, wracked by instability, portends increasing economic insecurity for all but the wealthiest 1% of the population. And with this upheaval on the climate and economic fronts, we are witnessing the failure of the political establishment to offer a viable energy alternative.

All of this sets the context for a book that describes the way forward for a revolutionary movement in energy, one that wrests control and ownership of energy resources out of the hands of the energy establishment—democratizing energy and making it a vital resource for advancing the environmental, economic, and social justice needs of our communities.

Difficult as this challenge already was, the November 2016 election made it even more so. If the goal of this book was compelling before the right-wing takeover of the federal government, it is even more urgent now. The desperate acts of the current federal government to breathe new life into a dying fossil fuel economy, foment racial intolerance, reassert U.S. military dominance, eliminate health and other protections, and otherwise bow to corporate interests underscore the need to strengthen and empower our communities. Community control of energy resources will be a critical aspect of resistance to the right-wing agenda.

As we demonstrate in this book, confronting race, racial discrimination, and racial oppression is central to developing a sustainable, decentralized energy alternative. Moreover, the leadership of people of color is key to building a powerful energy democracy movement.

Recent events put this perspective in sharp relief.

The first is the central role of racism in engineering the right-wing takeover of the federal government, all in the interest of the most extreme "rogue" fossil fuel sector, whose agenda is to lay waste to the ecosystem at all costs. The 1% used race-baiting tactics—both subtle and blatant, anti-Muslim, anti-black, anti-Latino, and anti-immigrant fear mongering—throughout the presidential campaign to achieve unprecedented control and dominance over the environmental and economic survival needs of the 99%. Racism in the service of human extinction.

The second is the historic struggle at Standing Rock to assert the rights of indigenous communities to clean water, taking on an oil industry hell-bent on intensifying the climate crisis. The struggle against the Dakota Pipeline was supported by Black Lives Matter, environmental justice organizations, and many others who understand how the fossil fuel economy targets people of color. More significantly, the struggle also drew the support of other organizations and individuals—from environmentalists, to unionists, to antiwar veterans—who see how their fate is tied to the fate of people of color. There is a growing and heartening recognition, not only within racial and ethnic groups, but among broad communities across America, that, in fact, not one of us is free—or safe—unless all of us are.

Accordingly, the subtitle of this book—Advancing Equity in Clean Energy Solutions—emphasizes the centrality of racial and economic justice to an energy democracy movement.

Now, more than ever.

Acknowledgments

We want to thank each of the contributors to this volume, as well as Ronnie Kweller at Emerald Cities Collaborative and Heather Boyer and the team at Island Press, for supporting this project. We also want to acknowledge the many activists and advocates, past and present, who shape the thinking and practice of our movement, as well as Favianna Rodriguez, whose artistic expression of that movement was incorporated into the cover design of this book.

Denise Fairchild and Al Weinrub, *April 27, 2017*

CHAPTER 1

Introduction

DENISE FAIRCHILD AND AL WEINRUB

If there is a reason for social movements to exist, it is not to accept dominant values as fixed and unchangeable but to offer other ways to live—to wage and win, a battle of cultural worldviews . . . laying out a vision that competes directly with the one on harrowing display, . . . one that resonates with the majority of people on the planet, that . . . we are not apart from nature but of it.

Naomi Klein, *This Changes Everything: Capitalism vs. the Climate*

What does it mean to get real about climate change and take back control over our energy resources? What energy alternatives represent real solutions to the economic and environmental crisis confronting our civilization?

While still in its formative stages, energy democracy, a growing current in the clean energy and climate resilience movement, is attempting to address these very questions. Energy democracy is rooted in the long-standing social and environmental justice movements and is a key component of the evolving economic democracy movement. It goes beyond the simplistic "transition to 100% renewables" framework to offer a deeper understanding of the cultural, political, economic, and social dimensions of the climate change problem.

This volume collects the converging perspectives, strategies, and practices of the emerging field of activism that defines energy democracy. It highlights the promising ideas and efforts of U.S.-based energy democracy advocates and practitioners. As opposed to the academic, scientific, and policy perspectives of mainstream environmental professionals, this book gives voice to community-based organizations and leaders active in the climate and clean energy struggle. Their perspectives differ radically from the mainstream environmental community about how to get real about climate change.

The growing energy democracy movement is more important now than ever. Climate and social justice advocates are entering a new, shocking reality. The United States government is abandoning its already weak commitment to reduce greenhouse gas emissions, as represented by the 2015 Paris Climate Agreement. The new federal administration is staffing its cabinet and agencies with operatives of the fossil fuel industry, opening the door to fossil fuel exports and transcontinental oil and gas pipelines, bringing back coal and extreme extraction, and gutting environmental regulations and the federal agencies that oversee them.

Energy democracy addresses these challenges by joining the environmental and climate movement with broader movements for social and economic justice in this country and around the world.

The Energy Imperative

A global energy war is under way. It is being waged on numerous fronts, with distinct battle lines. It's man versus nature; global North versus global South; fossil fuel versus clean energy; globalization versus local sovereignty; the powerful moneyed class versus low-income and indigenous communities and communities of color (the haves versus the have-nots); and, fundamentally, an extractive economy versus a regenerative economy.

The stakes are high for everyone. The health of the planet and whether humans will survive as a species will be determined by who emerges as the victors of the warring factions. Fortunately, a growing global consensus points to the need to move from a fossil fuel economy to a clean energy economy. The Consensus Project (theconsensusproject. com) reports the near unanimous (97%) consensus among climate scientists that the massive burning of gas, oil, and coal is having cataclysmic and cascading impacts on our atmosphere and climate, depleting Earth's natural resources, including its land, food, fresh water, and biodiversity. Extreme weather events—the incidence of torrential rains, floods, heat waves, droughts, and hurricanes—resulting from global warming further threaten human settlements, life, and property.

Such climate disruption finally propelled 195 world leaders to sign the 2015 Paris Climate Accord. The Accord is an acknowledgment that the fossil fuel economy is no longer sustainable, that climate change is happening, that it is human-induced, and that a global effort is needed to stem the greenhouse gas pollution that threatens human survival.

Yet despite the urgency of the climate challenge, the fossil fuel sector—concentrated in five supermajors: BP, Chevron, Conoco, ExxonMobil, and Shell—continues to debunk climate change and disdains worldwide concerns about the existential threat of extracting, transporting, and burning increasing amounts of dirtier and harder-to-get fossil fuels.[1] The Dakota Access and Keystone North American transcontinental pipelines are but two examples of the continued corporate drive to wreak havoc on our fragile ecosystem, ruining delicate aquifers, sovereign First Nation lands, farm communities, the oceans, and, of course, Earth's atmosphere. These climate and environmental impacts are particularly magnified and debilitating for low-income communities and communities of color that live closest to toxic sites; are disproportionately impacted by high incidences of asthmas, cancer, and rates of morbidity and mortality; and lack the financial resources to adapt to climate impacts.

The Path Forward: Democratizing Energy

In the face of this threat to survival, the battle lines have been clearly drawn between fossil fuel capitalism (the fossil fuel industry, its Wall Street backers, and its military enforcers) and those working to avert climate disaster.

"We need to view the fossil fuel industry in a new light," says climate activist Bill McKibben. "It has become a rogue industry, reckless like no other force on Earth. It is Public Enemy Number One to the survival of our planetary civilization."[2]

Many different forces are opposing the determined efforts of the fossil fuel industry to continue its program of globalization, extreme energy, and international military hegemony, at everyone else's (and Earth's) expense.

This opposition includes the struggles against fossil fuel extraction (the Keep It in the Ground movement), especially opposition to the further development of extreme energy (the Keystone Pipeline, deep ocean drilling, Arctic drilling, tar sands exploitation, hydrofracking, and so forth); opposition to fossil fuel subsidies; opposition to oil wars; regulation of carbon emissions; imposition of carbon taxes; shutting down coal-fired power plants; and other areas of struggle that unite diverse forces in opposition to the "rogue" fossil fuel capitalists.

In essence, this opposition is attempting to wrest control of energy resources from the powerful institutions that are driving humanity to the brink of extinction. The struggle reflects an effort by citizens to exercise more control over energy decisions and to self-determine a sustainable, life-supporting energy future.

While this opposition needs to be deepened and strengthened, there remains an important strategic question: what is the alternative to the fossil fuel energy–based global economic system?

A large number of climate activist organizations in the United States are engaged in efforts to contain the fossil fuel establishment's increasingly desperate program of extreme energy extraction, climate destabilization, and environmental destruction. Many of these organizations have the goal of reducing greenhouse gas emissions or even transitioning to a decarbonized energy system.

This resistance has awakened many people, politicized them around energy and climate issues, and fueled an increasingly powerful grassroots opposition to the corporate energy agenda. However, these movements are still mainly reactive and have exhibited, for the most part, only a limited vision of an energy alternative.

Many, for example, call simply for a technological fix: for a transition to 100% renewable energy, citing how it is technologically possible to develop sufficient renewable resources. But these calls do not specify who will develop and control that energy, to what end, or to whose benefit. The impetus is to decarbonize the economy, but otherwise leave the basic economic and social system—the institutional framework—intact.

This approach fails to confront the capitalist growth imperative that jeopardizes the world's ecosystem, or to address the globalized exploitation of human and natural resources that leaves billions of people struggling to survive, or to fully appreciate how climate disruption, gross economic disparities, oppression, and institutionalized racism are inextricably linked.

Naomi Klein, in her book *This Changes Everything: Capitalism vs. the Climate*, demonstrates that the climate crisis draws into question the institutions and logic that have created our existential predicament. She points out both the necessity and the opportunity of our thinking outside the box, creating truly *transformational* solutions, if we are to survive.

In this vein, a growing number of climate activists see resistance to the corporate energy agenda as a struggle for social, racial, environmental, and economic justice. These "climate justice" forces see the opposition to fossil fuel capitalism as a key front in a crucial battle to transform our economic system more deeply—an economic system that has used fossil fuel energy as the driver of capital accumulation, ecosystem destruction, and social exploitation. For these activists, the struggle *against* the extreme fossil fuel agenda is a struggle *for* system change, for an alternative system. It is a struggle *for* community health, community resilience, and community empowerment. It is a struggle for social justice and an opportunity for building community.

The struggle is not simply to decarbonize the economic system, but to *transform* it.

Hence, the question we posed above—the alternative to the fossil fuel energy–based global economic system, is a justice-based ("just") transition to a new, renewable energy–based, ecologically sound, equitable, life-sustaining economic system that can serve the needs of the world's peoples.[3]

And in case it is not obvious, let's be explicit. The struggle to achieve that kind of alternative is fundamentally a struggle for democracy.

The Energy Democracy Movement

An international labor roundtable[4] in October 2012 framed the struggle for a global energy transition as an issue of democracy: "An energy transition can only occur if there is a decisive shift in power toward workers, communities and the public—*energy democracy*. A transfer of resources, capital and infrastructure from private hands to a democratically controlled public sector will need to occur in order to ensure that a truly sustainable energy system is developed in the decades ahead. . . . "[5]

In short, energy democracy is a way to frame the international struggle of working people, low-income communities, and communities of color to take control of energy resources from the energy establishment and use those resources to empower their communities—literally (providing energy), economically, and politically. It means bringing energy resources under public or community ownership and/or control, a key aspect of the struggle for climate justice, as described earlier, and an essential step toward building a more just, equitable, sustainable, and resilient economy.

Thus, the energy democracy movement—represented by a growing number of organizations and organizing campaigns worldwide—seeks to replace our current corporate fossil fuel economy with one that puts racial, social, and economic justice at the forefront of the transition to a 100% renewable energy future.

In particular, energy democracy acknowledges the historical and contemporary perspectives and experiences of frontline communities— those most directly impacted by the fossil fuel economy and by the impacts of climate change, as well. This framing prioritizes the needs and concerns of working families, indigenous communities, and communities of color in the struggle to define a new energy future. It seeks comprehensive and effective solutions to the full impact of the fossil fuel economy.

Energy democracy is a critical framework for addressing the economic and racial inequalities that a decarbonized economic system would otherwise continue to perpetuate.

A New Energy Paradigm

The energy democracy movement implies a profound shift in how we think about and relate to energy. Energy is an essential enabler of all human activity—from producing the essentials of life, to transportation, to communication, to the creative arts. We can't survive without it. In that light, given the existential threat we now face as human beings from the burning of fossil fuels, our relationship to energy must be reevaluated; this involves a paradigm shift of major proportions. The new energy paradigm must address three major aspects of our energy system: its relationship to the environment, to social justice, and to a new economy.

A NEW ENVIRONMENTAL PARADIGM

Energy democracy represents a new environmental paradigm.

Energy democracy emphasizes the core values and related strategies needed to protect Earth's species. It seeks to find the historical and cultural precedents for making our energy systems life-sustaining, relying on ecological principles from preindustrial, traditional, and land-based societies.

Much is now understood by scientists about the impact of fossil fuel use on the environment. The extraction of these fuels is laying waste to huge tracts of land and ocean due to mountaintop removal, deep-water drilling, tar sands oil production, fracking, and other forms of extreme fossil fuel extraction. The burning of these fuels—increasing concentration of greenhouse gasses in the atmosphere and the acidification of the oceans—is modifying Earth's climate and altering the biosphere, causing the extinction of an ever-increasing number of species and now putting human populations in jeopardy as well.

This ecosystem destruction is a product of the industrial, fossil fuel economy, which has accelerated mass production and consumption and the accumulation of wealth. But its origins lie in a Western civilization worldview of human beings as masters and exploiters of the natural world for the betterment and progress of human civilization—without regard to the fragile ecosystem needed to maintain life on the planet or the delicate balances required for Earth to sustain life.

Energy democracy seeks to reframe energy from being a commodity that is commercially exploited to being a part of the commons, a natural resource to serve human needs, but in a way that respects the Earth and the ecosystem services provided by the biosphere. The new paradigm calls for reducing the human footprint, reducing waste, and reducing energy use as key to ecosystem health and stewardship. From this perspective, energy—both fossil fuel *and renewable*—is a communal resource requiring democratic ownership structures and sustainable, ecological management. This view runs counter to the commodification of energy that underlies many clean energy strategies today.

The ideas of the "commons" and "just transition" are discussed in this volume to present a different way to think about and value our environment and to provide a pathway from commodifying our energy assets toward democratizing them.

A NEW SOCIAL JUSTICE PARADIGM

Energy democracy represents a new social justice paradigm.

Energy democracy recognizes the racialized impacts of the fossil fuel economy and of climate change and sees them as threat multipliers: they deepen the daily economic, health, and social justice challenges of vulnerable communities. The new energy democracy paradigm harnesses the lived experiences of low-income communities and communities of color to reverse that impact and to design an alternative energy system.

The fossil fuel economy has had a disproportionate impact on people of color in the United States. The rise of fossil fuel power in the last two hundred years was a key factor in replacing the slave system of production with free labor and in industrializing and commercializing the U.S. economy. The result was the westward expansion, growth of urban centers, rise of monopoly capitalism, concentration of wealth, migration and immigration of working-class people and people of color, segregation, impoverishment, and creation of urban slums.

In the period following World War II and the dominant role of oil in suburban sprawl, many black and Latino communities were left to live in industrial zones, near toxic release sites and coal-burning power plants, as a result experiencing severe health impacts. For example, the burning of fossil fuels is accompanied by mold spores, dust, and particulate matter, decreased ozone protection, and toxic chemical pollutants that lead to respiratory ailments, cancers, heat-related morbidity and mortality, human development and mental and stress-related disorders, and vector- and water-borne morbidity and mortality.[6]

Moreover, federal and local land-use, housing, and transportation policies, along with bank redlining, trapped low-income populations in these toxic communities, giving rise to the U.S. environmental justice movement.

Beyond the more direct racialized impacts of the fossil fuel economy are the racial impacts of climate change itself. Those hit hardest by the extreme weather conditions induced by climate change—the floods, the droughts, the hurricanes—are communities of color.[7] Hurricane Katrina and Superstorm Sandy stand out as examples of how the poorest populations and neighborhoods were least prepared to withstand and recover from the impact of these storms, amplified by weak levees, inadequate energy infrastructure, contaminated water, and failed sewer and transportation systems.

Possibly more significant are the impacts of longer-term climate change, such as hotter weather, drought conditions, and extreme weather. These are damaging our agriculture sector and impacting food supply. Agricultural workers, largely part of low-income immigrant communities, will lose work and the ability to support their families. Shifts in the availability and price of agricultural products will make it particularly more difficult for low-income families to put food on the table.

Clearly, the entire human species is threatened by the environmental impacts of climate change. But the most vulnerable populations are least able to afford escalating prices of food and other necessities, lack access to health services, are last to receive emergency services, live in housing and communities most vulnerable to floods and heat waves, and lack financial resources to deal with the impacts of climate change or to bounce back from extreme weather events.

Energy democracy addresses these challenges by emphasizing the importance of building community resilience among the most vulnerable, with an emphasis on community-based renewable energy development in building the economic strength, resilience, wealth, and power of low-income communities and communities of color. It focuses on those most negatively impacted by the fossil fuel economy. It stresses equity and the need to redress historical harm in finding solutions that achieve social justice.

The new paradigm also recognizes and raises up the leadership of indigenous communities, low-income communities, and other historically marginalized communities in the struggle for climate justice and in envisioning and shaping an energy system that can meet the needs of all people. The contributors to this volume describe how their struggles to resist the harmful impacts of the fossil fuel economy help build an alternative, equitable energy system.

A NEW ECONOMIC PARADIGM

Energy democracy represents a new economic paradigm.

Energy democracy sees renewable energy resources as enabling a new, alternative economy—a regenerative rather than an extractive economy, one that builds the economic strength and resilience of our nation's communities. This new economy model is characterized by community-based development, nonexploitive forms of production, socialized capital, ecological use of natural resources, and sustainable economic relationships. It emphasizes the need for meaningful work and for family-sustaining jobs for all workers.

By contrast, our current economy, built on the back of fossil fuel energy, has achieved for vast increases in labor productivity by exploiting natural resources and human labor to accumulate capital and create huge corporate empires. The most affluent Americans (the 1%) are wealthier than ever, yet most of us continue to experience serious economic distress, and many communities, particularly communities of color, are, literally, in a state of economic crisis. Acute inequality and exclusion are becoming the new normal.

With the growth of a deregulated low-wage economy, maximizing profits and growth are the only operative standards. Big-business and large financial institutions have assumed state-like powers in ruling over workers, communities, and democracy itself. Under this economic regime, we have witnessed a chronic failure to create jobs; an increasingly regressive tax system (lower taxes for the wealthiest, higher taxes for everyone else); accelerating income inequality (not only increasing but increasing at an increasing rate); growing housing costs and increased homelessness; increasingly costly health care, defunding and privatization of the educational system, an eroding retirement system, growing corporate control over government and politics, and increasing surveillance of U.S. citizens and restrictions on our rights and civil liberties.[8]

We have also witnessed exponentially expanded material consumption over the last century and a half to where that consumption is now colliding with the material limits of the Earth's resources and its capacity for ecosystem restoration.

The fossil fuel economy is at the root of many of these global economic, social, and environmental dislocations. The industry receives massive public subsidies to fuel the overproduction and consumption of our natural resources, investing in extreme extraction and processing that produce mounting negative externalities. Researchers at the International Monetary Fund determined that the world's governments are providing subsidies to the highly profitable oil industry to the tune of an astonishing $5.3 trillion in benefits per year, including direct subsidies as well as the cost of health impacts, environmental spills, and other environmental calamities.[9] This represents $10 million every minute, every day.[10]

Simply decarbonizing the current economic system—hard as this might be—by transitioning to a nonfossil, renewable energy base does not challenge the fundamental logic or economic power relationships of this extractive global economy. It does not impact the growth imperative of the capitalist system nor stop Wall Street and the largest U.S. corporations from extracting wealth from working people. It does not address income and wealth inequality. Decarbonizing this economic system extends its life.

A regenerative, life-sustaining economic alternative, like any economic model, needs an energy model attuned to its values and needs. We cannot build a new economy on an old energy model. Energy democracy—

bringing energy resources under public or community ownership and control or other forms of cooperative economics—promotes a more regenerative, equitable, sustainable, and resilient economic system.

This volume presents varying views and models for advancing energy democracy, including municipalizing energy utilities, reclaiming ownership and control of rural electric co-ops, and creating various implementations of the decentralized energy model.

The Need for a New Energy Model

The paradigm shift represented by energy democracy is more than a new set of values and principles to guide our energy system. The paradigm shift calls for a new energy model, one in sync with the environmental, social justice, and new economy paradigms described above.

The dominant renewable energy model today—the *centralized* renewable energy model—is an extension of the legacy model of fossil fuel electrical energy production to renewable energy. This model represents corporate control of the energy system. However, an alternative model— the *decentralized* renewable energy model—allows for control and ownership of renewable energy resources to reside in the community, rather than in remote corporate boardrooms. This decentralized energy model represents an opportunity for community empowerment.

CENTRALIZED RENEWABLE ENERGY

The centralized renewable energy model is based on large-scale centralized generating systems—big solar plantations and large wind farms— which are the product of concentrated financial and economic power. Only on rare occasions are centralized energy developments the result of democratic action of communities, and in such cases, a centralized energy system can provide economic wealth building and political democracy for those communities. However, in most cases, centralized energy development represents the interests of powerful economic forces aided by a corporate state apparatus unfettered by democratic restraints.[11]

Centralized renewable energy is the model of choice for a corporate decarbonized economic growth strategy and its drive for continued capital accumulation. That strategy emphasizes a transition to industrial-scale, carbon-free energy resources without challenging the growth of energy consumption, material consumption, rates of capital accumulation, and concentration of wealth and power in the hands of a few.

DECENTRALIZED RENEWABLE ENERGY

By contrast, the decentralized renewable energy model enables community-based renewable energy development. It allows for the new economic and ecologically sound relationships needed to address the current economic and climate crisis.

Thus the decentralized renewable energy model involves not only the shift from fossil fuel power to renewable power but also the shift from corporate control of energy systems to more democratically controlled energy systems. Democratic control of renewable energy resources, in particular, is facilitated by the fact that renewable resources are *distributed*: solar energy, wind, geothermal energy, energy conservation, energy efficiency, energy storage, and demand response systems are resources that can be found and developed in all communities. These distributed energy resources provide a basis for community-based *decentralized* development of energy resources at the local level through popular initiatives.[12]

The decentralized renewable energy model represents a global energy realignment that has begun to gain traction around the world as people struggle to take control of renewable energy resources.

Advancing Energy Democracy: A Map of the Book

Initiatives to transform the large-scale, corporate-controlled dirty energy economy into a community-owned and controlled clean energy economy are emerging throughout the United States.

These energy democracy initiatives include communities that are participating in planning, building, and benefiting from an alternative, decentralized renewable energy model. These initiatives focus on community control of, access to, and ownership of energy assets. They intentionally address the health and economic conditions of communities most impacted by fossil fuel energy and include strategies for workers and communities impacted by the downsizing of the fossil fuel industry.

In essence, this energy democracy movement is a comprehensive climate resilience initiative to address the existential consequences of the extractive economy through the creation of a new regenerative economy, one based on a decentralized renewable energy model that advances ecosystem health, economic sustainability, and social justice through the empowerment of our communities, and the democratization of our society. It is situated within, and makes intentional linkages with, the broader social justice and new economy movements.

In this book, we present some of the most important initiatives, organizations, and leaders pushing forward the energy democracy movement in the United States.

The contributors discuss their perspectives and approaches to climate and clean energy from the lens of rural Mississippi, the South Bronx, and California immigrant and refugee communities to urban and semirural communities in the Northeast. The racial, cultural, and generational perspectives represented in this volume are as pronounced as the geographic ones. This diversity is bound together, however, by a common operating frame: that the global fight to save the planet—to conserve and restore our natural resources to be all-life sustaining—requires an intersectional approach. It must fully engage the community and must change the larger economy to be sustainable, democratic, and just. In essence, there is a growing kinship and unity between climate justice and the other movements for human, civil, worker, immigrant, and democratic rights.

The contributions span a range of perspectives, from strategic to practical, focusing on policy frameworks, organizing strategies, community engagement, and energy development models—all various expressions of what it takes to build a viable energy democracy movement.

This volume knits together the disparate threads of this movement to help crystallize the problem; codify the values, principles, and strategies of energy democracy; and curate the possibilities for change through information, knowledge, and models offered by a number of energy democracy leaders.

A number of the contributors refer to the values and principles that guide their work. Because the contexts are different, the different guiding principles might seem inconsistent. But while they vary in nuance, all share a set of fundamental values: respect for the Earth, ecosystem diversity, sustainable stewardship, racial justice, reversing and repairing historical harm, economic justice, community ownership and control of energy assets, self-determination, and community governance. These are the values and principles that constitute the paradigm shift of energy democracy.

Each chapter stands alone, however, offering insights into energy democracy from different cultural perspectives, communities, and strategies.

Cecilia Martinez from the Center for Earth, Energy and Democracy (CEED) suggests that energy democracy is a human rights issue and examines the indigenous notion of the "commons" as a cultural frame for rebuilding our economy into one in which natural and human resources are respected, protected, and equitably shared. Cecilia offers the relevant overarching principles and contrasts them with the dominant paradigm of the extractive economy.

The chapter on a just transition, by Michelle Mascarenhas-Swan of Movement Generation, highlights the work of the Our Power Campaign, which recognizes the ecosystem as the operating center for all solutions related to climate change, the economy, and social equity. Michelle describes how community-based organizing brings us back to basics—home—and advises us to be mindful of the family and economic aspects of transitioning from fossil fuels to community clean energy. It is critical not only to address the economic burdens of change for the current energy labor force but also to ensure that the new, regenerative energy economy adequately supports family-sustaining employment.

Strela Cervas, of the California Environmental Justice Alliance, and Anthony Giancatarino, until recently of the Center for Social Inclusion, focus on a state policy framework for moving financial and technical resources into historically disadvantaged or overburdened communities ("just transition" or "green" zones). These communities are sites for developing decentralized renewable energy projects and other new economic institutions that can transform economic life for those communities. This approach has been very successful in California in achieving legislation to shift the needed resources to such programs.

Vivian Yi Huang and Miya Yoshitani of the Asian Pacific Environmental Network (APEN) discuss the community base building and leadership development of their organization to build the political power needed to achieve the local programs and legislative successes in California. Vivian and Miya show how energy democracy is not a top-down enterprise, but one that is rooted in the experience, capacity, voice, and power of local resident immigrant communities. They demonstrate how APEN is building knowledge, leadership and power among their constituents to be key stakeholders in energy policy and planning.

Derrick Johnson and Ashura Lewis of One Voice, in Mississippi, focus on the organizing work taking place in rural southern communities to reclaim community ownership of existing energy cooperatives—the rural electric co-ops—and to transform them into authentic community-owned and controlled energy service providers. The rural electric co-ops are key to the economic revitalization of low-income communities in the South, and this work has strong implications for the hundreds of rural electric co-ops around the country.

Meanwhile, Sean Sweeney of Trade Unions for Energy Democracy (TUED) points out that the interests of organized labor are in conflict with the economic underpinnings of the extractive economy—including its emphasis on the private accumulation of wealth, exploitation of labor, and income and power disparities—and how that conflict has created an energy democracy current within the international labor movement. Sean explores what this means in the U.S. context, where we see an explicit fracturing of the trade union movement, with a few unions now moving toward a strong energy democracy stance.

Al Weinrub, of the Local Clean Energy Alliance, addresses the democratizing of energy at a municipal scale, in particular through Community Choice energy, a vehicle available in a number of states.

Community Choice enables local jurisdictions to form public energy service-provider agencies, somewhat akin to municipal utilities. The possibilities for democratized, decentralized energy are huge, but, to date, the potential of Community Choice has not been fully realized. Al describes the strategies and organizing under way to ensure community control, equity, local wealth creation, and other community economic benefits possible through Community Choice.

Maggie Tishman, of the Bronx Cooperative Development Initiative, highlights the strategy and practice of using anchor institutions to finance decentralized energy projects to improve the health, local economy, and resilience of low-income communities and communities of color in the Bronx. Maggie describes how the procurement and investment capital of anchor institutions is mobilized for community-based energy projects that improve a community's physical, economic, employment, and environmental health by supporting development of local start-ups, social enterprises, and cooperatives. She also describes the institutional barriers that must be overcome.

Lynn Benander, Isaac Baker, and Diego Angarita Horowitz, of Co-op Power, explore the strategic vision and core values of energy cooperatives and the ways they have overcome the many barriers involved. They describe a new kind of energy cooperative, one representing an alternative economic development model for advancing a new renewable energy economy. They present case studies that highlight the economic, social, and environmental benefits being made available to low-income communities through these new energy cooperatives.

Finally, Anya Schoolman and Ben Delman, of the Community Power Network, highlight community-based solar projects to further community ownership of renewable energy resources. They argue that this model builds a political base of support needed to create new policy, transform the utility sector, and resist efforts of the monopoly utilities to undermine the emerging clean energy movement. Their vision is of a future in which every community participates in the financial benefits of our energy grid's ongoing shift from a big, centralized energy system to a distributed system of locally owned, decentralized community energy resources.

The book closes with an admonition that a powerful energy democracy movement is not easy, quick, or assured. Denise Fairchild, of the Emerald Cities Collaborative, posits that the vision and movement for a fundamentally different energy system and society rocks the status quo—upsetting a legacy of entrenched power, privilege, property, and profits. Denise suggests that the seemingly unfathomable task of dismantling the fossil fuel economy is comparable to what it took to dismantle the slave economy: it was a long struggle, brought the United States into civil war, involved a global struggle, and required amending the U.S. Constitution. Denise concludes that just as the abolitionist movement ended the formal institution of slavery, the energy democracy movement can end the fossil fuel economy, advancing a more humane and just economic system—one that protects both our natural and human resources and ensures our long-term survival.

Taken together, the contributions in this volume provide the backbone for reengineering our energy system to be not only carbon-free, but just. They represent different efforts to create decentralized energy systems that can empower our communities. They plant the seeds for a larger movement for structural change.

We hope, by example, that the contributions in this book show what an alternative, democratized energy future can look like and inspire others to take up the struggle to join the emerging energy democracy movement.

1. Dana Nuccitelli, "Facts Matter and on Climate Change, Trumps Picks Get Them Wrong," *The Guardian*, accessed December 27, 2016, https://www.theguardian.com/environment/climate-consensus-97-per-cent/2016/dec/27/facts-matter-and-on-climate-change-trumps-picks-get-them-wrong.
2. Bill McKibben, "Global Warming's Terrifying New Math," *Rolling Stone*, July 19, 2012, http://www.rollingstone.com/politics/news/global-warmings-terrifying-new-math-20120719.
3. Al Weinrub, *Labor's Stake in Decentralized Energy*, p. 4, http://energydemocracyinitiative.org/wp-content/uploads/2012/10/Labors-Stake_10-22-121.pdf.
4. See http://energydemocracyinitiative.org.
5. Sean Sweeney, *Resist, Reclaim, Restructure: Unions and the Struggle for Energy Democracy*, October 2012, Executive Summary, http://energydemocracyinitiative.org/required-reading-roundtable-discussion-document.

6. National Institute of Environmental Health Services, www.niehs.nih.gov/research /programs/geh/climatechange/health_impacts/weather_related_morbidity/index.cfm.

7. Rachel Morello-Frosch, Manuel Pastor, James Sadd, and Seth B. Shonkoff, *The Climate Gap: Executive Summary—Inequalities in How Climate Change Hurts Americans and How to Close the Gap* (2010).

8. Jack Rasmus, "America's Ten Crises," June 30, 2012, http://www.kyklosproductions.com /posts/index.php?p=160.

9. Rmuse, "Report Shows the Oil Industry Benefits from $5.3 Trillion in Subsidies Annually," *Politics USA*, June 9, 2015, http://www.politicususa.com/tag/annual-oil-subsidies.

10. Ibid.

11. Two days after the historic 2014 climate march in New York City calling for climate action, federal and California State officials released an 8,000-page proposal for private renewable energy development on 22.5 million acres of California desert. See Carolyn Lochhead, *Energy Plan Calls for Big Renewables Projects in State's Deserts,* September 23, 2014, http://www.sfgate.com/green/article/Sprawling-solar-farms-OKd-near-desert-national -5775871.php.

12. A ten-minute video in which Al Weinrub describes decentralized energy systems and their community benefits can be found at http://www.youtube.com/watch?v=HvuXxyKSh3A.

CHAPTER 2

From Commodification to the Commons:
Charting the Pathway for Energy Democracy

CECILIA MARTINEZ

There was a time when people thought of the Earth's environment as durable and impervious to the manipulations and activities of human society. That was until knowledge about the existence of environmental problems such as acid rain, ozone depletion, ocean acidification, and climate change led to an awareness that humans impact the Earth's natural systems. In a span of less than three hundred years, we have managed to seriously impair the quality of our air and water, alter the Earth's climate systems, and craft the beginning of the planet's sixth mass extinction.[1] These problems are extraordinarily complex and can overwhelm the desire and capacity to act. But unless we design and implement solutions we will continue on course to fundamentally alter nature, exposing communities around the globe to great harm. Thus, there is clearly an urgency to act.

But how do we design effective solutions to address these problems in a timely manner? There are the climate deniers who refute any need for action at all. Thankfully, there are many people who respect climate science and recognize the need for change, but there are competing ideas about the course of action. Some believe the best approach is to invest in technological innovations that will result in a greener or more environmentally sustainable development path with continued economic growth. Others argue that political and economic systems that continue to produce a development path that is inherently unsustainable need to be changed. Additional perspectives fall within this continuum and often result in conflicting strategies for how to move forward.

The Centrality of Energy

One thing that environmentally conscious parties can agree on is the importance of the energy system in our current environmental predicament. The United States relies on fossil fuels for four-fifths of its energy needs,[2] and internationally fossil fuels make up 81% of the world's energy consumption.[3]

The electricity sector alone, notes Benjamin Sovacool, director of the Danish Center for Energy Technology, "is so big that it consists of almost 20,000 power plants, half a million miles of high-voltage transmission lines, 1,300 coal mines, 410 underground natural gas storage fields, and 125 nuclear waste storage facilities, in addition to hundreds of millions of transformers, distribution points, electric motors, and electric appliances. It is the most capital intensive sector of economic activity for the country and represents about 10% of sunk investment."[4]

Beyond the electricity system, the entire energy sector includes an extractive economy that operates hundreds of thousands of mines, wells, drilling platforms, and pipelines to supply energy for our current daily needs. The national demand for crude oil alone averaged 19.4 million barrels *per day* in 2015,[5] while the world demand of 96 million barrels of oil and liquid fuels each day is expected to reach the 100 million barrels-per-day milestone by 2021.[6]

The scale and scope of the global fossil fuel economy has increased the carbon content in the atmosphere so much that it now threatens our oceans and fresh water sources, food supply, and the overall viability of our ecosystem. But to confine our examination of the fossil fuel energy system solely to environmental problems is incomplete and insufficient. The energy system has also had a significant role in producing conditions of economic and social inequality.

Over the last two hundred years, our society has built an energy infrastructure based on the idea that social progress is dependent upon a steady expansion of energy production. That is, the social and economic inequality and environmental costs were inconsequential when compared with the alternative of a stagnant economy, and the imperative of economic growth was dependent upon a constantly expanding energy supply. This development model was supported by a worldview that has come to value the environment as just another commodity.

Unfortunately, the problem of social inequality that resulted from this development path remains relatively unexplored in domestic energy policy. Sovacool and Dworkin, authors of the 2014 book *Global Energy Justice*, note that it "is becoming increasingly clear that routine energy analyses do not offer suitable answers to these sorts of issues. The enduring questions they provoke involve aspects of equity and morality that are seldom explicit in contemporary energy planning and analysis."[7]

We are in a moment of history that requires an alternative energy path that addresses the formidable environmental *and* social issues that threaten our future. On the positive side, we are moving toward a clean energy economy. A significant number of public and private initiatives have been developed and deployed for this transition. A range of state and local policies promoting and incentivizing renewable energy production have been implemented. Federal regulations on the transportation and power sectors are in place, and corporate sustainability initiatives are becoming more prevalent.

Despite these modest gains toward a clean energy future, a more radical and holistic approach is needed, especially in light of the 2016 electoral outcomes. There are now efforts to aggressively repeal environmental protections, suspend actions to address climate change, and undermine basic rights and guarantees. Consequently, because of its integral role in society, a new energy pathway must assume a proactive role in advancing an alternative paradigm that not only fixes our environment, but other equally pressing social issues as immigration, living-wage jobs, criminal justice and policing, and civil and human rights.

This new energy agenda must be rooted in the principles of democracy and justice first. It must embrace a new understanding of the role of energy in society—one that acknowledges that the natural resources that are transformed into usable energy are part of the Earth's commons. Shifting the paradigm from "energy as commodity" to "energy as commons" will be fundamental to achieving a sustainable, just, and democratic future.

This chapter (1) explores why a domestic energy policy that can move us toward long-term sustainability must also embrace the concepts of democracy and justice, and (2) provides guiding principles for creating energy democracy.

Energy as a Commodity in Carboniferous Capitalism

Reminiscent of the times, University of Michigan professor Leslie A. White argued in his 1943 article "Energy and the Evolution of Culture" that cultural advancement was a product of the technical capability of a society to use energy to transform the materials of nature into "human need–serving goods and services."[8] History, he argued, could be divided into three distinct phases: a period of "savagery" in which human survival depended upon reaping what was available in nature through hunting and gathering; a period of "barbarism" in which agriculture and animal husbandry were developed; and a period of engines and fuels, which marked the beginning point of "civilization." Among the disturbing issues with this perspective is that social progress is intrinsically tied to the ability of a society to use energy for the mass extraction and consumption of natural resources.

Over the course of its development, the energy system produced staggering growth. Between 1920 and 1950, energy consumption doubled from 19.8 to 39.7 quadrillion BTUs (British thermal units), doubled again between 1950 and 1975, and reached a peak of 101 quadrillion BTUs in 2007.[9] By 2015, total U.S. energy consumption decreased to 97 quadrillion BTUs,[10] in part due to greater efficiencies. Electricity consumption also grew phenomenally, doubling each decade during the postwar period. This energy diet helped to triple the nation's gross national product (GNP) in the first half of the twentieth century.[11]

The extraction, processing, and transmission of fossil fuel–based energy have resulted in a multitrillion-dollar enterprise. ExxonMobil, the largest U.S. fossil fuel corporation, generated $259.5 billion in revenue and $71.2 billion in profits in 2015.[12] In a span of less than three hundred years, the industrial economy has supported the mass extraction and consumption of natural resources to meet its energy demand.

The present fossil fuel energy system could not have been possible without the underlying social tenet exemplified by White that energy abundance was, and is, integral to social progress. The political and economic agenda throughout the twentieth century required decisions not only to legitimize but also to intentionally promote corporate energy organizations as the vehicle for securing this abundance. This meant that the millennia of natural processes embodied in coal, natural gas, and petroleum were now treated as resources that not only could be, but *should be*, extracted for inputs into the industrial economy.

Criticisms or objections to this energy development path over the years was, and continues to be, greeted with the retort that to incorporate these environmental costs will undermine economic growth. A classic example of this is former president Nixon's declaration of the need for maintaining the fossil fuel energy complex at a time when the country was experiencing severe energy price increases and restrained supply due to the 1970s oil embargo, as well as major environmental threats such as ozone depletion, acid rain, and oil spill disasters. In speaking to the seafarers union, he stated: "We use 30% of all the energy . . . that isn't bad; that is good. That means that we are the richest, strongest people in the world and that we have the highest standard of living in the world. That is why we need so much energy and may it always be that way."[13]

As former chairman of the Federal Reserve Board Ben Bernanke stated before the Economic Club of Chicago on June 15, 2006, "At the most basic level, oil and natural gas are just primary commodities, like tin, rubber or iron ore. Yet energy commodities are special, in part because they are critical inputs to a very wide variety of processes of modern economies. They provide the fuel that drives our transportation system, heats our homes, and offices, and powers our factories."[14]

This view of "energy as commodity" is prevalent today even as the energy industry transitions to renewable energy sources. Renewable energy has evolved from an alternative grounded in a movement for a different social and political future to fast becoming a big business pushed by new investments, technologies, consumer demand, and regulatory requirements. The result is that planned grid-connected solar photovoltaics—nearly nonexistent in the recent past—totaled 9.5 gigawatts (GW) in 2016, and wind capacity additions totaled 6.8 GW of utility generating power.[15] Even distributed solar—that is, small-scale photovoltaics (PV) at the residential and commercial scale—is dependent upon the economic and technological largess of corporate energy, as its viability is tied to various incentive programs that allow owners to sell their surplus power back to the grid.

The clean energy industry, therefore, remains one firmly ensconced in the current energy regime.[16] Four major questions guide public and private decisions about energy choices: (1) Can the energy technology become market viable, which essentially means profitable? (2) Can the technology "fit" or potentially "fit" the existing energy infrastructure? (3) Can the technology achieve a scale that matches or mimics a highly centralized corporate form of operations and management (the latest preoccupation often discussed as taking a strategy to scale)? (4) Due to the urgency of climate change, can the technology reduce carbon emissions? Except for the carbon reduction criterion, wave power, wind power, solar power, hydropower, nuclear power, carbon capture and sequestration, and "clean coal" technology have all the above factors in common with fossil fuels.

Unequal and Unsustainable: The Legacy of Energy Commodification

The social, economic, and political processes that supported the development of the energy-as-commodity infrastructure have also produced and reinforced patterns of racial and economic inequality.

At the international level, the energy inequality relationship is seen in the extreme development and wealth disparities between the global South and North. The North cannot escape the fact that it has attained its prosperity by using the Earth's natural resources. At the same time, the development of industrial economies over the last two centuries has been the main contributor to climate change. According to the United Nations 2007 *Human Development Report*, wealthy industrialized nations are responsible for "about 7 out of every 10 tonnes of carbon dioxide (CO_2) that have been emitted since the start of the industrial era."[17]

As Anil Agarwal and Sunita Narain, two leading researchers and activists from India, state, global warming occurs in an unequal world.[18] The United States generates a gross domestic product of approximately $17 trillion annually and emits on average per capita CO_2 emissions of between 17 and 18 metric tonnes per year. In contrast, those countries on the bottom of the UN Human Development Index emit an annual average of less than 0.1 metric tonnes of CO_2 per capita.

The human and environmental costs of this resource-intensive development are borne disproportionately across communities. In nearly every sector, from housing and transportation to forestry and mining, the expansive growth of the fossil fuel–powered economy exploits people as well as the environment. Commodification of human beings enabled acceptance of slave and low-wage labor; commodification of land led to appropriation, clearing, and mining of native homelands; and the use of nature as a depository for waste and pollution were all consequences of commodification.[19] As the politics of energy policy continues to unfold, it is important not to mask the historical realities of unequal development that have led us to the present day.

The question is, How can we move forward on an energy transition that addresses inequality and environmental degradation? The answer lies in recognizing that these two phenomena are linked—and achieving one without the other is neither desirable nor effective. True sustainability requires the integration of justice and equity, which is accomplished by developing a democratic ethic. Energy democracy is not merely a matter of instituting more meaningful processes of community engagement in an inherently undemocratic system, or of providing affordable and accessible energy regardless of the social and environmental costs associated with the life cycle of its source. Bullkeley, Edwards, and Fuller, three geography professors from the United Kingdom and Australia, noted in a 2014 paper on climate justice that it is "important to establish whether interventions in the name of climate change serve to maintain the interests of an elite at the expense of a minority, and as such perpetuate patterns of inequality . . . or whether they are instead able to shift the terms of debate, make space for alternatives, and address existing forms of inequality."[20]

Fighting for Democracy: Creating an Energy Commons

It is no longer tolerable or affordable to ignore the human and environmental costs that were an accepted and institutionalized part of the fossil fuel economy. So how are we to move forward? How can we effectively, efficiently, and urgently act to transition to a sustainable energy future? Pursuing a sustainable energy future requires moving from the energy-as-commodity regime to "energy as commons."

In the international arena, the concept of the global commons refers to areas of resources that exist outside the political reach of any one nation. Thus, international law identifies four global commons: the high seas, the atmosphere, Antarctica, and outer space. However, human history shows that there is a diversity of ways in which communities and cultures have devised commons-type governance to organize and share resources. As Nobel laureate Elinor Ostrom stated, "Communities of individuals have relied on institutions resembling neither the state nor the market to govern some resource systems with reasonable degrees of success over long periods of time."[21]

I offer that the first step in instituting energy commons governance is to recognize that energy is, in reality, the transformation of a vast array of natural interactions and phenomena for societal use. Whether the energy source is a fossil fuel, such as coal and oil, or derives from a renewable process, such as solar irradiation or wind, society is using these natural endowments to meet its needs. A fundamental principle of the energy-as-commons approach is that these natural endowments should not be owned by, or belong to, any set of peoples, countries, or corporations exclusively. Nor should any one generation assume the right to overexploit or exhaust these resources. Further, the organizations that extract, refine, transform, and transmit energy should operate as a democratic system.

Next, rules and institutions that govern energy-related organizations and activities need to be created to support the commons. Beyond the contemporary versions of the high seas and outer space, there are numerous examples of communities developing commons governance systems that align with principles of democracy and justice.

Customary cultural practices by peoples in the Southwest continue to operate and maintain this type of governance over water, a highly contentious and scarce resource in the region. The *acequia system* of water allocation depends on common maintenance of a network of earthen ditches and on a collective commitment to the principle of sharing water in times of plenty and scarcity. It includes networks for the exchange of labor and resources in order to provide an intricate ecosystem of mutual support for small-scale farms and the community. The acequias are designed so that both individual and community can flourish.[22] This form of water governance is the legacy of hundreds of years of customary practices among indigenous people in the United States. There are similar examples in communities around the world, including India and the Middle East.

Indigenous governance can serve as an important starting point for practical lessons on the energy transition. Sustainability scholars at the College of Menominee Nation document how "sustainability models used throughout the world can be problematic for indigenous communities because they do not often address or incorporate indigenous cultural values, concerns, worldviews (epistemologies and ontologies) or teachings. Indigenous concepts often excluded from sustainability models include reciprocity (mutual responsibilities guiding human and non-human interactions), interrelationships among humans and non-humans (all things are related), cooperation, and respect."[23]

In juxtaposition to the norms of commodification, markets, and material abundance associated with the present energy system, indigenous worldviews tend to share a common value of reciprocity. Reciprocity is the recognition that the well-being of human society cannot be divorced from the well-being of the environment; a healthy environment means a healthy human community, and a healthy human community supports a healthy environment.[24] Human communities are important, but their importance is "tempered by the thought that they are dependent on everything in creation for their existence."[25] Indigenous values and practices offer examples of commons governance. In addition to Native tribes and community initiatives, numerous community-oriented energy strategies across the United States reflect a similar orientation.

Moving Toward the Energy Commons

Energy as commons offers a gateway to transition from the corporate, centralized fossil fuel economy to one that is democratically governed, designed on the principle of no harm to the environment, and is supportive of diverse local economies that provide sustainable livelihoods for all community members. This energy transition is a long-term endeavor, which includes redeveloping planning and policy mechanisms that explicitly incorporate and support these goals. However, getting from here to there also requires very practical solutions. Following are a set of useful guidelines for assessing whether the proposed policies move us toward an energy commons or, alternatively, whether they reinforce the present energy system.

- **Align with the human rights framework:** Energy is a vital and basic need and is essential for the quality of life. It is internationally recognized that a safe, clean, healthy, and sustainable *environment* is an essential precondition to the full enjoyment of the wide range of *human rights*, including the rights to life, health, food, and water.[26] To be in alignment with a human rights framework, acknowledging the interrelationship between human rights and the environment is insufficient. Energy governance institutions must establish mechanisms for effective, full, and equal environmental protection and energy planning. Such institutions include federal and state agencies responsible for pollution regulation; public utilities and service commissions (PUCs and PSCs); energy agencies such as the Federal Energy Regulatory Commission (FERC), the North American Electric Reliability Corporation (NERC), the Nuclear Regulatory Commission (NRC); regional transmission organizations (RTOs) and independent system operators (ISOs); and federal and state legislative bodies. There must be a right to information, spaces for effective and meaningful participation in decision making, and mechanisms for redress. As former EPA administrator Lisa Jackson has suggested, "We can't on the one hand say that a clean environment is a basic right, and not say that it is the right for everyone."[27]

Under a human rights framework, there is an obligation and responsibility to ensure that community members are participants in designing the architecture of the energy system. Because the energy system consists of national and international supply chains, the character of energy production and consumption in one community has direct and indirect impacts on other communities. Thus a principle of the commons is that all peoples should have the right and ability to meaningfully participate in, and make decisions about, the energy structure and the costs and benefits impacting their community. Meaningful participation by community members requires that public institutions provide them with the information and tools to do so. Yves Lador, the permanent representative of EarthJustice to the United Nations in Geneva, states:

Individuals cannot rely just on their own direct access to the environment to foresee these threats. They need institutions to tell them what air they are breathing, or what water they can drink. The actions required for one's livelihood, which are individual ones by nature, can only come now from collective action. It is therefore a necessity for individuals to have access to information in order to take personal decisions and to be able to take their responsibilities. Thus the importance of the right to know and of an understanding of how decisions are made. . . . The relation between States and their citizens must also progress towards more trust, more transparency and more participation."[28]

- **View energy as service.** In the same way that the food democracy movement emphasizes a renewed understanding and relationship to food, energy democracy promotes a relationship to nature that is embedded in an ethic based on sustainable stewardship. This new relationship emphasizes that the natural resources that produce the energy to nourish our economic needs have intrinsic value beyond commodified inputs. The way food is produced and consumed reflects our values and relationship to the plants and animals that provide for our sustenance and to the communities and peoples that grow, harvest, transport, and produce the food. Similarly, in our relationship to energy we must include knowledge and recognition of how these resources came into being for our use and the costs embodied in their availability.

This perspective fundamentally differs from the prevailing view of energy, in which: (1) the role of a community member is reduced to that of consumer; (2) energy used for luxury items is not differentiated from energy used for basic living; and (3) little attention is given to the normative or ethical aspects of energy consumption.

There is evidence of some progress in transitioning to an energy services approach. The role of energy efficiency (reducing the inefficiencies or wastefulness of consumption) has increased substantially in recent years. However, there are also indications of co-optation of the energy services concept for the benefit of maintaining the present corporate energy structure. To use one example, energy efficiency programs remain the domain of the utility sector, often only because they are mandated in order to reduce wasteful energy consumption. While this is a step in a positive direction, the question is, To what degree are such efforts designed to keep the current corporate structure in place, albeit more accountable, and are there other mechanisms for energy efficiency that are more effective and democratic that may stand in conflict with utility-based efficiency delivery?

The energy transition requires an understanding that producing energy for consumption and profitability is not the end, but the means to providing services essential for quality of life for all community members.

- **Plan at an appropriate scale and for diversity.** Energy as commons requires community-based energy planning projects in which community members themselves are active planning agents about the type of energy infrastructure they envision for themselves that can best meet community needs. Energy alternatives such as distributed energy (e.g., rooftop solar or community solar) and other on-site renewables and energy efficiency options can be effectively developed and implemented at community scales. Moreover, the responsibility for decisions regarding acceptable environmental risks and protections becomes a functional part of the energy commons through democratic local governance structures.

  This approach to energy planning runs counter to the present energy system, which is preoccupied with developing renewable and community energy options that are large scale and can be incorporated into the existing centralized energy supply infrastructure. The justification for these options is often the sunk costs in the central grid, the need for reliability during peak hour demands or to provide redundancy (backup) to minimize the effects of electricity disruption and blackouts. They do not, however, consider democratic principles and decision-making processes.

  Our ultimate goal should be to create energy options that embody commons principles, reduce inequalities, and support human rights for communities across the life cycle of energy production and consumption that is best achieved through community control of decisions.

- **Acknowledge the rights of nature.** Nature is a dynamic living system comprised of unknown numbers of living systems and organisms—each interrelated and interdependent. This principle refers to the responsibility of human society to acknowledge and respect the integrity of living systems, including the conditions for regeneration and protections against destruction and degradation.

  The indigenous concept of justice that applies to all life (human and nonhuman) helps to provide for a more holistic or comprehensive understanding of human-environmental relations. Whether human beings demonstrate wisdom and reverence for the natural world is a matter of choice, but ultimately all life, even human life, cannot escape accountability for our choices and actions. As a tribal member stated: "Did you know that trees talk? Well they do. They talk to each other, and they'll talk to you if you listen. Trouble is, white people don't listen. . . . But I have learned a lot from trees; sometimes about the weather, sometimes about animals, sometimes about the Great Spirit."[29]

Modern culture has detached itself from nature to such an extent that environmental degradation has become a normalized part of individual and collective life. The dominant view is that plants, animals, and the biological and physical processes of the Earth are objects for human exploitation. However, there is another way to understand our role within the planetary ecosystem. Drawing from indigenous knowledge and practice, a more respectful approach to society's relationship with nature is available.

Creating a Viable Energy Democracy

Structurally, the environmental crisis and the crisis of democracy are the same problem and thus must be solved together.[30] While it may seem overwhelming at times, we should not lose sight of the fact that we can make better choices for the future. Ultimately, energy democracy requires acknowledgment of the history of uneven impacts and environmental destruction that have resulted because of the fossil fuel complex. It will be a challenge to reconstruct the current energy system, but there is opportunity to create a viable energy democracy by reengaging the commons in principle and in practice.

The energy-as-commons approach offers the possibility to transition out of an unhealthy and unsustainable consumerist relationship to one in which people and communities understand and accept responsibility for the consequences and impacts of their energy consumption. This challenges clean energy advocates to expand the concept of energy democracy beyond a conventional understanding of representative democracy. Energy democracy is not simply creating community energy projects that continue to operate within the same practices of commodifying nature or under the same corporate-based structure. Rather, actions that empower and community members and provide them with opportunities to redefine their relationship to the environment are core. The task before us is to define how energy democracy will guide our actions over the next century.

1. Center for Biological Diversity, *Elements of Biodiversity: What and Where It Is*, accessed December 29, 2016, www.biologicaldiversity.org/programs/biodiversity/elements_of_biodiversity/
2. U.S. Energy Information Agency, 2016, *Frequently Asked Questions*, March 17, accessed January 8, 2017, https://www.eia.gov/tools/faqs/faq.cfm?id=33&t=6.
3. World Bank, 2014, *Fossil Fuel Energy Consumption*, accessed 11 13, 2016, data.worldbank.org/indicator/EG.USE.COMM.FOZS?end=2013&start=1960&view=chart.
4. Benjamin K. Sovacool, *Energy and Ethics: Justice and the Global Energy Challenge* (New York: Palgrave Macmillan, 2013).
5. U.S. Energy Information Agency, 2016, *Frequently Asked Questions*, March 17, accessed January 8, 2017, https://www.eia.gov/tools/faqs/faq.cfm?id=33&t=6.
6. International Energy Agency, 2017, *Oil*, accessed January 8, 2017, https://www.iea.org/about/faqs/oil/.
7. Benjamin K. Sovacool and Michael H. Dworkin, *Global Energy Justice: Problems, Principles, and Practices* (Cambridge: Cambridge Univerisity Press, 2014).
8. Leslie A. White, "Energy and the Evolution of Culture," *American Anthropologist* 45 (1943): 335–56.
9. U.S. Energy Information Agency, *Frequently Asked Questions*, March 17, 2016, accessed January 8, 2017, https://www.eia.gov/tools/faqs/faq.cfm?id=33&t=6.
10. Ibid.
11. Sam H. Schurr and Bruce Carlton Netschert *Energy in the American Economy, 1850–1975: An Economic Study of Its History And Prospects* (Baltimore: Johns Hopkins University Press, 1960).
12 . Statista, 2017, *ExxonMobil's Revenue from 2001 to 2015*, accessed January 5, 2017, http://www.statista.com/statistics/264119/revenue-of-exxon-mobile-since=2002/.
13. Richard M. Nixon, cited in Sargent Shriver, Speech to the National Catholic Education Association, April 25, 1975, accessed November 30, 2016, www.sargentshriver.org/speech-article/speech-to-the-national-catholic-education-association.
14. Ben S. Bernanke, Energy and the Economy, Speech to the Economic Club of Chicago, June 15, 2006, accessed February 16, 2017, https://www.federalreserve.gov/newsevents/speech/bernanke20060615a.htm.
15. U.S. Energy Information Agency, *Frequently Asked Questions*, March 17, 2016, accessed January 8, 2017, https://www.eia.gov/tools/faqs/faq.cfm?id=33&t=6.
16. Leigh Glover, "From Love-Ins to Logos: Charting the Demise of Renewable Energy as a Social Movement," in *Transforming Power: Energy, Environment, and Society in Conflict*, edited by John Byrne and Noah Toly, pp. 247–68 (New Brunswick, NJ: Transaction Publishers, 2006).

17. United Nations Development Programme, *Human Development Report 2007/2008: Fighting Climate Change in a Divided World* (New York: Palgrave Macmillan, 2007).

18. Arun Argawal and Sunita Narain, *Global Warming in an Unequal World: A Case for Environmental Colonialism* (New Delhi: Centre for Science and Environment, 1991).

19. Cecilia Martinez, "Energy, Technics and Postindustrial Society" (diss., University of Delaware, 1990).

20. Harriet Bulkeley, Garreth A. S. Edwards, and Sarah Fuller, "Contesting Climate Justice in the City: Examining Politics and Practice in Urban Climate Change Experiments," *Science Direct* 25 (2014): 31–40.

21. Elinor Ostrom, *Governing the Commons: The Evolution of Institutions for Collective Action* (New York: Cambridge University Press, 2015).

22. Gregory Allen Hicks and Devon G. Pena, "Community Acequias in Colorado's Rio Culebra Watershed: A Customary Commons in the Domain of Prior Appropriation," *University of Colorado Law Review* 74, no. 2 (2003): 387–486.

23. Michael J. Dockry, Katherine Hall, William Van Lopik, and Christopher M, Caldwell, "Sustainable Development Education, Practice, and Research: An Indigenous Model of Sustainable Development at the College of Menominee Nation, Keshena, WI, USA," *Sustainability Science* 11, no. 1 (2016): 127–38.

24. Ibid.

25. Vine Deloria Jr., *God Is Red* (Golden, CO: Fulcrum Publishing, 1994).

26. United Nations Development Programme, *Human Development Report 2007/2008* (New York: Palgrave Macmillan, 2007).

27. Rosenthall/Jackson interview, July 6, 2009, https://www.youtube.com/watch?v=xZBde8KNYXU.

28. Yves Lador, "The Challenges of Human Environmental Rights," in *Human Rights and the Environment: Proceedings of a Geneva Environment Network Roundtable*, pp. 7–13 (Geneva: United Nations Environment Programme, 2004).

29. Vine Deloria Jr., *God Is Red*.

30. John Byrne, Steven M. Hoffman, and Cecilia Martinez, eds., *The Social Structure of Nature, Proceedings of the Sixth Annual Meeting of the National Association of Science, Technology and Society, 1991*, pp. 67–76.

CHAPTER 3

The Case for a Just Transition

MICHELLE MASCARENHAS-SWAN

In 2009, Movement Generation Justice & Ecology Project [1] co-led a delegation of U.S. grassroots groups to the UN Climate Change Conference in Copenhagen. Appalled by the failure of the "Big Greens" to address the root causes of the climate crisis, in 2010, these groups brought others in to form what would later become the Climate Justice Alliance. The purpose was to join forces to leverage the power these community-based organizations had been exerting to stop or reduce the harm from mining, refining, transporting, and power generation operations in their own communities and tribal lands.

To scale up their power to stop the growth of the dirty energy-fueled economy, grassroots community organizations would unite across the chain of destruction and chain of custody of those operations, while at the same time providing critical leadership for a transition to just and regenerative economies. In 2012, the Climate Justice Alliance officially launched the Our Power Campaign to unite communities on the front lines of the struggle around cultivating such a "just transition."

Just transition is a framework for a fair shift to an economy that is ecologically sustainable and is equitable and just for all its members. After centuries of global plunder, the profit-driven, growth-dependent industrial economy is severely undermining the life support systems of the planet. An economy based on extracting resources from a *finite ecosystem faster than the capacity of the system to regenerate* will eventually come to an end—either through collapse or intentional reorganization. *Transition is inevitable. Justice is not.*

Just transition strategies were first forged by labor unions and environmental justice groups who saw the need to phase out the industries that were harming workers, community health, and the planet while also providing just pathways for workers to transition to other jobs. In the 1990s, an organization called the Just Transition Alliance, a coalition of environmental justice and labor organizations, began bringing workers in polluting industries together with "fenceline communities" in groundbreaking conversations that forged a new understanding of economy and home. Building on that history, just transition has come to mean forging coordinated strategies to transition whole communities toward thriving economies within their control and that provide dignified, productive and sustainable livelihoods, democratic governance, and ecological resilience.

As illustrated in figure 3-1, a just transition requires shifting from dirty energy to clean community power, from building highways to expanding public transit, from incinerators and landfills to zero waste, from industrial food systems to food sovereignty, from gentrification to community land rights, and from rampant destructive development to ecosystem restoration. Core to a just transition is deep democracy in which workers and communities have control over the decisions that affect their daily lives. Constructing a visionary economy for life calls for strategies that democratize, decentralize and diversify economies (and energy) while damping down consumption, and, through reparations, redistributing resources and power.[2]

Our Just Transition Framework

Our just transition strategy framework asserts an integral relationship between economy, energy, and equity.

WHAT WE MEAN BY ECONOMY

Since the 1950s, we in the United States have been taught to equate the word *economy* with gross domestic product, the rise and fall of the stock markets, the unemployment rate, and other measures that relate exclu-

Figure 3-1. Local, living, and loving economies for life *Image source: Movement Generation Justice & Ecology Project*

sively to the success or failure of capitalism to increase the accumulation of profit and the creation of jobs.

But the root of the word *economy* is *eco*. *Eco* comes from the Greek word *oikos*, which means "home." So *economy* is most simply defined as the *management of home*. Home is nested in a web of relationships that can be defined as an *ecosystem*. And though the current dominant extractive economy has tried to disentangle the economy from the environment, the escalation of climate disruption makes it clear that the economy is not separate from the environment but is, instead, rooted firmly in it.

If economy means management of home, how that management

occurs is key to the outcomes. If economic decisions are made by people or forces far from where the impacts of those decisions are felt, the result is a likely mismanagement. As indigenous peoples across the globe have long understood, we can only manage home well if we understand the impacts we have on the place where we are rooted. That requires *ecology*, or *knowledge of home*. There is no "one size fits all" on planet Earth. To ensure sustainability for seven generations, to use the measure advised by the Iroquois Nation, human communities will identify unique ways to meet their needs given the ecosystem they depend on. One way to think of an ecosystem is the "basin of relations." This basin of relations defines the watershed, foodshed, energyshed, and tradeshed of an economy.[3]

While there are an untold number of ways of managing home—different types of economies—every economy is woven from a common set of threads (figure 3-2). We take *natural resources* and combine them with *human labor* (a particularly precious natural resource) toward some *purpose*. The purpose of an economy can be the accumulation of wealth and power, or the purpose can be meeting needs toward the sustainability of future generations in a place.

Every economy is rooted in a culture or *worldview* that makes the particular form of economy make sense to the people who participate in it. The culture or worldview—the songs, stories, languages, practices, rituals—interacts with the other threads of economy, influencing and being influenced by them. Finally, there is *governance*, which literally means "to steer." Through governance, groups create the rules, norms, activities, and accountability mechanisms needed to make the economy function as smoothly as possible toward its purpose.[4]

THE POLITICAL ECONOMY OF ENERGY

As the editors note in the introductory chapter, "Energy is an essential enabler of all human activity." Energy provides the ability to do work.

All living things on Earth use energy originating from the sun to transform the material world through their work. Bees pollinate as they gather nectar, creating fruit. Soil microorganisms convert dead matter into compost and, after millions of years, fossil fuels. The com-

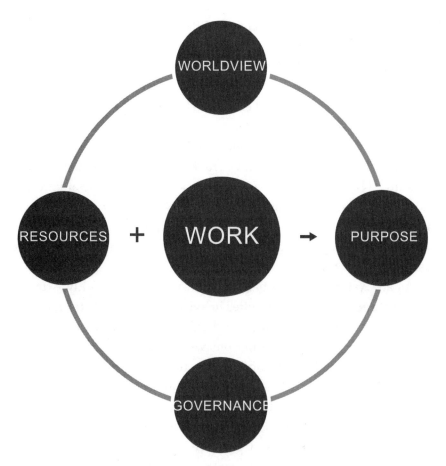

Figure 3-2. The threads of economy *Image source: Movement Generation Justice & Ecology Project*

bined work of everything from fungi to redwoods, from microbes to polar bears, and from plankton to people creates the material basis upon which human communities—in fact all life on Earth—depends. Human labor can be applied to take these forms of wealth—or resources—and convert them into more wealth. Human labor can save seeds or build soil. Or, as in the case of the United States, which was powered by the labor of enslaved African people during the first 250 years of its burgeoning economy, human labor can be applied violently to extract more from an ecosystem than it returns.

41

Our ability as humans to harness different forms of energy—coal, oil, natural gas, geothermal, nuclear fission, or even food to power human work—depends on our access to the resources needed to do so. Fossil fuel energy, for example, requires large amounts of capital to drill for oil, or to remove mountaintops to get at coal, or to transport these fuels, or to build large electricity-generating plants.

In the United States, the capital needed was first accumulated from the plantation economy, the slave trade itself, and later, the industrial economy.[5] Nuclear fission as an energy source, first for the atomic bomb and later for nuclear power reactors, is made viable by decades of capital accumulation and concentration. This capital was the product of an extractive economy that consistently took more resources from the ecosystem than it returned.

As an example, the globalized food system is a highly extractive economic sector. This sector relies heavily on fossil fuels for fertilizers, herbicides, pesticides, tractors, and transport, as well as on landless workers and precarious populations who need jobs to put food on the table. While the industrial food system has not solved the problem of hunger in the world, it has created an abundance of other problems that now threaten individual human health, public health, and the life support systems of the planet. The global food system uses about ten times more energy than it produces; contributes nearly one-third of greenhouse gas emissions; creates costly public health problems, including diabetes and heart disease; and is responsible for topsoil loss, water pollution, and a waste stream that is often burned or landfilled, further exacerbating climate change and public health problems. This problematic food system is the result of peoples stolen from or pushed off their lands, stripping them of the resources required to produce food in ways that align with short- and long-term ecological and social well-being.

Similarly, if we look at the electricity and transportation fuels sectors, we find an extractive economy that leaves the vast majority of U.S. people almost completely powerless to impact the ways they generate, access, or use energy. The extraction and burning of fossil fuels is taking the planet to the brink of many climate tipping points and creating populations who are increasingly unable to manage home in a historical period that requires more ingenuity, creativity, thoughtfulness, rootedness, and leadership to pivot quickly to avoid an even more catastrophic and dystopic future.

According to the California Energy Commission, "Water-related energy use consumes 19 percent of the state's electricity, 30 percent of its natural gas, and 88 billion gallons of diesel fuel every year."[6] Working to shift the management of water (and related energy use) to communities stewarding their watersheds requires neighborhood and community-level design and installation workshops to create rainwater catchment and greywater sys-

tems that "slow it, spread it, and sink it" rather than "pave it, pipe it, and pump it."[7] Working with plumbers and pipefitters unions, labor can be applied at the community level to create pathways for new water technicians adapting age-old technology to retrofit water systems that meet community needs while restoring watersheds, in the process cutting the use of fossil fuels and related greenhouse gas emissions (figure 3-3).

The increasing concentration and control of resources—land, water, energy (including human labor), and more—has eroded the capacity of human communities to

Figure 3-3. Nineteen percent of California electricity is used to pump water. Participants in Movement Generation and Occidental Arts and Ecology's Permaculture for the People training program design and install a rainwater catchment system to utilize gravity-fed flow. *Image source: Brock Dolman*

manage home well. In the extractive economy, we are almost completely unable to apply our labor to the living world around us, note the impact, and make decisions about how to further apply our labor. This means that we are almost completely unable to govern our own labor in ways that build resilience.[8] It is this most precious energy resource—human labor—that must be restored to democratic control in order to address the climate crisis and the array of related crises of the extractive economy.

Social Equity: Why Frontline Communities Must Lead the Transition

To manage home well requires us to apply ecology—the knowledge of home. This means observing and understanding how our actions impact the living world we depend on. Managing home well requires that communities have control over the decisions about how energy—in all its forms—is harnessed and applied. In fact, social inequity is a form of ecosystem imbalance. It will inevitably lead to the erosion of our ability to read and respond to our own impacts on ecosystems.

Through the industrial era, black, indigenous, Latino, Asian, Pacific Islander, and working-class white communities in the United States have experienced a disproportionate share of harmful impacts of the extractive economy, with the fossil fuel, nuclear, and waste incineration sectors being a key driver of the industrial economy.[9] Since the purpose of corporations is to maximize profit and most corporate decision makers do not live in the places they are impacting, corporate decisions rarely take into account the consequences of their actions on ecosystems, including human communities.

The result is a litany of fenceline communities harmed by corporate activities that pollute the air or groundwater, generate toxic waste, or are unsafe for workers or community members. For the past several decades, communities on the front lines of the impacts have been organizing to stop these bad actors from harming the places where they live, work, play, and pray. They form the backbone of the Environmental Justice movement that has asserted the adage, "no decisions about us without us."

Many of these place-based community groups have successfully stopped or reduced fossil fuel and other industrial harms through organizing community members to expose the damage and make business as usual untenable. These campaigns have cut the pollution impacts in environmental justice communities while simultaneously reducing greenhouse gas emissions at the source.[10]

Thus, these "fenceline communities" have found themselves on the front lines of the climate crisis. They are not alone. More and more communities are now organizing themselves from the many "front lines" of the root causes, impacts of, and false solutions to climate disruption. From those impacted by gentrification and speculative land ownership to fossil fuel infrastructure projects to false promises of green energy solutions, rural and urban communities are increasingly on the front lines of taking on the extractive economy that is undermining home.

In order to foster a just transition, these organized frontline groups increasingly understand that their communities must lead with their own solutions. In fact, who leads the transition will determine how it goes and where it lands.

To take one example, despite the impactful campaigns of environmental justice groups across the country, when these communities have not also put forth their own alternative economic program, time and again, extraction has continued to drive development. Sometimes it was toward expanded fossil fuel infrastructure and sometimes toward promises of "green energy solutions" such as nuclear power, algae-based biofuels, or waste incineration. These environmental justice groups understand that such "solutions" are false promises if they either exacerbate or cause new social inequity or ecosystem disruption.[11] Frontline communities have a crucial leadership role to play in ensuring that the inevitable economic transition is just—they have a firsthand understanding that justice requires that no community be a sacrifice zone. This lesson can be seen in the history of Native peoples offering refuge to black people who had escaped from slavery to form maroon societies. It can be seen in the role of women to end gender oppression. It can be seen in the role of free black people risking their own freedom to form the backbone of the Underground Railroad and the abolitionist movement. It can be seen in the role that queer, transgender, and gender-oppressed people play in overturning a gender-binary system that is harmful to all, though some more than others.

THE SPECIAL ROLE OF LABOR

It is the exploitation of workers that enables the extraction of natural resources from the ecosystem in a way that degrades rather than supports regeneration. Whether it is blowing up mountains for coal or slashing down forests for timber, the labor of working people is literally used as a chain saw against the very web of life they depend on.

A just transition requires that human labor be organized through democratic and voluntary cooperation, rather than coercion and exploitation. When they have the resources to make their own labor productive, communities across the globe resist blowing up mountains or building open-pit mines. Self-determined, embodied human labor—complete with awareness, feelings, instincts, thoughtful engagement, and the ability to act on those—is a key energy resource in regenerative economies. Along with other renewable resources, it must replace the destructive energy systems of the extractive economy. It is workers themselves who are best positioned to organize around this understanding. Every day, we see more and more organized labor groups finding their front lines—from teachers and cafeteria workers to nurses and an array of service workers. This is a critical and growing base of leadership for a just transition.

Strategies for a Just Transition

Shifting to managing home in concert with the principles of living systems requires aligning three core strategies: *building the new, changing the rules,* and *moving the resources* from an extractive to a regenerative economy. All of this must be done in a way that *stops the bad* and *changes the story.* Working with the Our Power Campaign, Movement Generation crafted figure 3-4 to illustrate a strategy framework to shift from an extractive economy to regenerative economies. The just transition strategy framework gives a way to understand the underlying threads of the current extractive economy and the envisioned threads weaving regenerative economies together as referenced in figure 3-2.

Fostering just transition requires building a movement of movements by aligning workers, environmental justice communities,

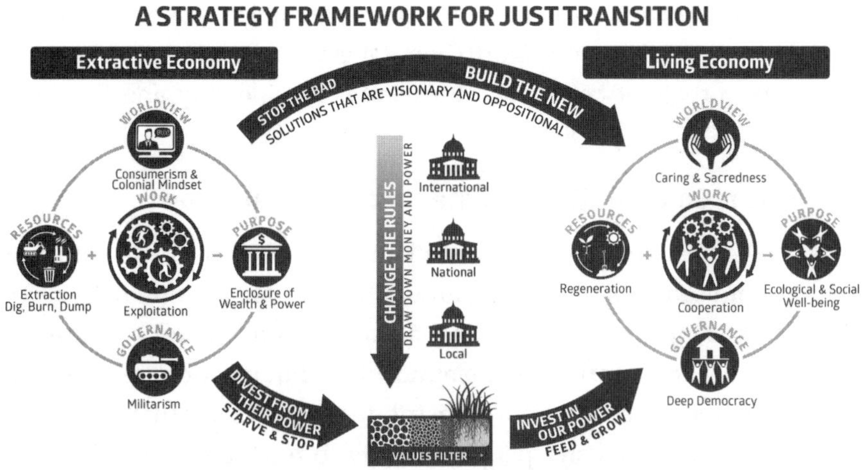

A STRATEGY FRAMEWORK FOR JUST TRANSITION

Figure 3-4. Just Transition Strategy Framework with the Our Power campaign *Image source: Movement Generation Justice & Ecology Project*

tenant rights groups, environmentalists, faith groups, students, resilience practitioners, indigenous peoples, black communities, queer and transgender communities, and countless others who have been harmed by the extractive economy. As these groups place their struggles into a framework of transition, the harms they have suffered under an extractive economy can be transformed into their leadership and vision for a new path that puts people and the places we depend on at the center. This creates a force of people, organized through social movements, who are actively putting their hearts, labor, and creativity toward building economic infrastructure, applying their own labor to meet needs rather than relying on the extractive economy to meet those needs. This economic infrastructure builds economic power that can be translated and applied as political power to further push rule changes and resource shifts toward local, living, loving, linked economies.

Building off the work of Dr. Vandana Shiva and others, a just transition calls for reorganizing economic activity using five key principles. Solutions that move us away from extraction and domination and toward cooperation and caring must *diversify, democratize, decentralize, reduce con-*

sumption, and *provide reparations.* An energy democracy movement must hold these principles as our North Star if it is to foster energy systems that do not exacerbate or create new forms of social inequity and the consequential ecosystem erosion.

Our Power: Communities Uniting for a Just Transition

In 2012, the Climate Justice Alliance launched the Our Power campaign to unite communities organizing around a just transition (table 3-1). The Our Power campaign holds that climate disruption will not be addressed simply by reducing carbon in the atmosphere but by addressing the root causes: the imbalance in resource distribution and power that leads to the erosion of ecosystems and people's ability to read and respond to their impacts. Our Power seeks to shift out of centralized and undemocratic energy systems that rely on fossil fuels, nuclear power, megadams, and waste incineration to renewable energy, including human labor. Our Power seeks to shift decision making from its current concentration far from its impacts to deep democracy — in which place-based peoples manage home toward a long-term view of future sustainability.

Our Power seeks to scale transition not by creating larger and larger organizations with greater concentrated power, but by uniting across issues and communities on a larger shared project of linking regenerative economies. Like a bag of marbles adding up to the weight of a bowling ball, this aggregation creates scale. While solutions will be applied locally, communities' ability to wrest control of economy from the current governing forces requires these local communities to band together in ways that build movement muscle. Instead of national campaigns, Our Power calls for *translocal* movement building: autonomous place-based organizing that is connected across communities through a unifying vision, shared strategies, and common frames.

Our Power has been steadily uniting a growing set of communities to leverage their collective power toward transition. In the first three years of the campaign, seven communities served as pilot sites for fostering a just transition. Anchored by grassroots organizations that had been resisting the impacts of the extractive energy system on their communities, they sought

Table 3-1—From old power to our power

	Old Power	Our Power
Energy resources and technologies	Human labor applied in ways that allow greater extraction	Liberated human labor (hands, heart, instinct, body, mind)
	Oil, coal, natural gas, nuclear fission, waste incineration, mega hydro, internal combustion, fossil fuel-based fertilizers, waste incineration (petroleum-based plastics usually required as fuel)	Other animal sources (labor and small-scale biogas digestion of waste)
		Passive solar, photovoltaic solar, wind, geothermal, mini hydro; reduced use, and other appropriate-scale approaches
Decision makers	Corporate executives, Wall Street financial brokers, bankers, large-scale land owners and developers, the 1%	People rooted in place making decisions about how they acquire and use resources, including how they apply their labor
Geography	Concentration of power: corporate headquarters, investment banks, nation states	Governance both local and bioregional, depending on type of "shed": foodshed, energyshed, tradeshed
	Globalized transport of resources, people, and goods	Most production and trade at the bioregional scale.
		People for the most part rooted in place

to deepen their focus on the real solutions their members envisioned. These grassroots groups have built coalitions with an array of other partners to align their work around a vision, strategy, and narrative framework that can engage everyday people in building new economic infrastructure, changing the rules, and moving resources toward a just transition.

CASE STUDY: RICHMOND, CALIFORNIA[12]

In Richmond, California, the behemoth Chevron Corporation's oil refinery stands as the number one greenhouse gas emitter in the state. For decades, it ran the city as a kind of company town with all or most city council members making decisions in accordance with the oil company's wishes. Community members from organizations such as Asian Pacific Environmental Network and Communities for a Better Environment organized to make the refinery safer, to cut the pollution coming from the plant, to stop the expansion of the refinery to refine dirtier grades of crude oil, and more.

A decade ago, community members began to organize through the Richmond Progressive Alliance to run candidates for local office who would prioritize values of community health and well-being. They succeeded in electing progressive candidates to local office who were not beholden to the oil company. Groups like Urban Tilth began to acquire public land to grow food for the community, training, employing, and developing the leadership of local residents on those lands. Local residents founded Rich City Rides, the city's only bike shop, and began organizing community bike rides and bike repair activities to shift from car culture to bike culture.

Richmond groups have come together to lead Our Power Richmond. The goal is to work toward shifting transit, food, and energy systems toward renewables and people power rather than fossil fuels and corporate control. Together with other partners, Our Power Richmond organizes community members around a vision of a just transition.

In 2016, several of these organizations launched Cooperation Richmond, a revolving loan fund and cooperative incubator. Cooperation Richmond now organizes community members to put capital to productive use by applying their labor to build cooperative businesses that help foster a regenerative economy. This provides critical infrastructure to "build the new." Through translocal organizing across the country, Cooperation Richmond is part of a larger initiative called Reinvest in Our Power. This initiative has two primary arenas. The first is to form a "financial cooperative," or network of local loan funds that support communities that have been excluded from access to capital to build and run cooperative businesses. The second arena is to connect divestment campaigns to reinvestment pathways. In this way, communities can shift capital away from fossil fuels, prisons, and war industries and invest it in community-run vehicles for economic transition.[13]

Figure 3-5. At Rich City Rides bike shop, residents learn to repair bicycles and engage in bicycle access advocacy. *Image source: Najari Smith*

At the same time, groups with a constituency in Richmond, Asian Pacific Environmental Network, Communities for a Better Environment, and the statewide California Environmental Justice Alliance are advancing policy at the state level to draw down public resources to communities on the front lines of the climate crisis, such as Richmond (see chapter 4).

While "building the new" and "moving the resources" are key strategies for a just transition, a just transition also requires changing the current rules that favor the extractive economy. Without a new set of rules based on values of a regenerative economy, cities like Richmond are beholden to industry, big-box retailers, or market-rate housing developers to grow their tax base. With out-of-the-box thinking and progressive organizing, over the last decade, Richmond groups have asserted new rules that put local residents and community health first, going up against the goliath corporations that try to stop them at every turn.

For example, in 2008, residents passed Ballot Measure T taxing the value of raw materials used in the manufacturing process of Richmond industries.[14] Though it was struck down by the courts, the process of garnering community support for such a radical shift in policy led the Chevron refinery to strike a deal with the city council to pay the equivalent of a standard utility user tax. This has brought an additional $15 million annually to the city.

While the company claimed that the high tax rates in Richmond made it appealing to move its operations to one of its other refineries, city council member Jeff Ritterman was quoted in the local news as saying, "I say to those other cities, 'Up your tax rates.' All cities are fighting for their survival right now and, as a society, we need to take some of that profit. It shouldn't all be going into private business and the wealthy."[15] A growing just transition movement can help create pathways for this $15 million revenue stream to be put to use developing renewable energy, cooperative businesses, community-scale transportation, and more.

In another illustration of changing the rules, the Alliance of Californians for Community Empowerment worked with the City of Richmond to create Richmond CARES in 2012. This program was meant to advance the power of local residents over mortgage bankers. The program allowed the city to declare eminent domain of underwater mortgages in order to restore equity and prevent blight from foreclosures.[16] Mortgage bankers quickly responded, flexing their muscle to downgrade the city's bond rating and the city backed off. At the time, other California cities were considering passing similar policies. Organizing translocally around such a policy could help to increase the power of the collective localities to take on powerful banks and investment houses that put profit over community well-being.

Richmond sits at the crossroads of an array of strategies for transition. While frontline forces in Our Power Richmond are forging a *just* transition as described above, heavily capitalized industries continue to seek to extract wealth from the city. The oil giant, Chevron, has continued to pursue its plans to expand its oil refinery operations there. Biotech companies collaborating with the U.S. Department of Energy and the University of California at Berkeley have pursued plans to build a laboratory designed for research in synthetic biology. While the laboratory has been put on hold, the refinery expansion continues, along with an expansion of fossil fuel infrastructure in surrounding communities. The thrust of this expansion is to refine dirtier grades of crude oil coming from tar sands extraction or hydraulic fracturing in the Bakken Shale Formation in North Dakota. Through translocal organizing, Richmond communities have found common cause with communities on the front lines of tar sands in Alberta, Canada, and in fracking-impacted communities.

SNAPSHOTS: OTHER OUR POWER COMMUNITIES

Among the diverse set of Our Power communities are black, Latino, Asian immigrant, and multiracial urban communities as well as predominantly white rural and Native communities. They span the West Coast to the Midwest and the Southwest to Appalachia and the black-belt South. In addition to Richmond, California, the seven current Our Power Communities include San Antonio, Texas; Jackson, Mississippi; San Francisco, California; Eastern Kentucky; Detroit, Michigan; and Black Mesa, Navajo Nation. While each community holds a set of shared values and principles of a just transition, each is applying a diverse set of just transition strategies based on the unique needs and conditions of home.

On the Black Mesa plateau in Arizona and New Mexico, Navajo and Hopi people have been working to foster a just transition from an economy dependent on the extraction of coal and water to an economy powered by solar energy and, ultimately, to a restorative economy rooted in traditional land-based life ways. As a cofounder of the Our Power Campaign, the Black Mesa Water Coalition (BMWC) was instrumental in ending the use of ancient aquifer water for coal transportation by Peabody Western Coal in 2005. BMWC is currently working to utilize solar power owned by the Navajo tribe and communities to fund a long-term plan for transition to traditional land-based livelihoods. Says BMWC cofounder and board member, Enei Begaye-Peter, "A green economy is nothing new to indigenous peoples. We have been practicing this way of life in harmony with Mother Earth before there was a Wall Street. But today, what we strive to do is unite the modern nonpolluting technologies, such as wind and solar, with the traditional technologies, such as weaving and farming; and with that unity we can open up new doors of opportunity for ALL our people—young and old, college-educated and land-educated alike."[17]

In eastern Kentucky, Kentuckians for the Commonwealth (KFTC) and partners such as the Mountain Association for Community Economic Development have been working to transition away from mountaintop removal and the dominance of "King Coal" to a just transition to "a fair economy, a healthy environment, new safe energy, and an honest democracy."[18] KFTC has been working with electric cooperatives to innovate on-bill financing of clean energy and energy efficiency projects such as How$martKY. And as part of their work to expand democratic engagement so that community members are making the decisions that affect them, KFTC members are working to make electric cooperatives more open, fair, and transparent. At the same time, KFTC organizes statewide to change rules and draw down resources toward people-powered initiatives such as these.

In Detroit, an array of organizers, cultural workers, and other leaders are advancing a shift from dependence on the Big Three automobile manufacturers and the industrial energy system they required to a thriving local economy rooted in commons around water and food systems; health and healing; media and culture; zero waste; whole-child education; and other systems of meeting community needs. The East Michigan Environmental Action Council anchors Our Power in Detroit, and its building, the Cass Corridor Commons, serves as a nexus of community-led and just transition activity, from youth media production to clothing swaps to political education.

Just Transition Means Remaking Economy

What you do to the land, you do to the people. And what you do to the people, you do to the land.[19] The concentration and control of land and resources in the hands of a few, including the energy needed to grow food, harvest water, generate heat, build shelter, and more has resulted in ecological destruction: undermining human communities' ability to meet their needs in ways that sustain the places they depend on.

The consequences are dire: climate destabilization, rising seas, resource wars, and the collapse of the biological and cultural diversity upon which our collective well-being depends. This situation demands a just transition through which we realign the purpose of the economy with the healing powers of Mother Earth.

A just transition calls for fundamentally remaking economy in ways that advance *ecological restoration, community resilience, and social equity.* Through *ecological restoration*, place-based communities engage the full dimensions of their own human labor to protect and advance biocultural diversity—taking action in ways that are fully embodied with awareness of the world around us; creativity to solve complex challenges; instincts; and love of people and place. By creating the conditions to maintain that biocultural diversity in the face of ecosystem disruption, we foster *community resilience.*[20] Finally, through redistributing resources and power, we advance *social equity* and, consequently, restore the reflective, responsive relationship to place required for ecological restoration.

1. Movement Generation Justice & Ecology Project is a movement support organization that provides training and strategy facilitation and tools to advance a just transition. The line of inquiry and language of this chapter were developed by the organization's staff collective in dialogue with hundreds of organizations.
2. See Vandana Shiva, *Earth Democracy* (Cambridge, MA: South End Press, 2005).
3. Brock Dolman, *Basins of Relations: A Citizen's Guide to Protecting and Restoring Our Watersheds* (Occidental, CA: Water Institute, 2008).
4. Movement Generation, *From Banks and Tanks to Caring and Cooperation* (Oakland, CA: Movement Generation, 2017), accessed January 28, 2017, http://movementgeneration.org/justtransition/.
5. Howard Dodson, "How Slavery Helped Build a World Economy" (February 3, 2003), accessed January 19, 2017, http://news.nationalgeographic.com/news/2003/01/0131_030203_jubilee2.html.
6. California Energy Commission, "California's Water-Energy Relationship." (Sacramento, CA: CEC, 2005).
7. Brock Dolman, PowerPoint presentation, May 2014, Occidental Arts and Ecology Center.
8. For a more complete description of what we mean by resilience, see Movement Generation, *Redefining Resilience: Principles, Practices and Pathways*, accessed January 28, 2017, http://pathways-2-resilience.org/ebook/part-ii-redefining-resilience.

9. For references related to environmental justice, see the Energy Justice Network website at http://www.ejnet.org/ej/, accessed January 1, 2017.

10. "Letter from the Grassroots to One Sky," accessed January 4, 2017, http://grist.org/article /2010-10-23-open-letter-to-1-sky-from-the-grassroots/.

11. For a discussion of false solutions, see Rising Tide North America and Carbon Trade Watch, "Hoodwinked in the Hothouse: False Solutions to Climate Change," accessed 1/17/2017, https://ecology.iww.org/PDF/RTNA/HoodwinkedV2ENG_screen.pdf.

12. Movement Generation is based in the San Francisco Bay Area, less than ten miles from Richmond, California. The just transition strategy framework is strongly informed by the movements that have been forging a new path in Richmond. For that reason, the Richmond case study is lengthier than the others.

13. See http://www.ourpowercampaign.org/reinvest and http://www.theworkingworld.org.

14. https://ballotpedia.org/Richmond_Business_License_Tax,_Measure_T_(November_2008), accessed January 17, 2017.

15. Alexa Vaughn, "End Chevron's Perk Campaign to Start Next Week," *Richmond Confidential,* January 7, 2010, accessed January 17, 2017, http://richmondconfidential.org/2010/01/07 /end-chevrons-perk-campaign-to-start-next-week.

16. Shaila Dewan, "Eminent Domain: A Long Shot Against Blight," *NY Times,* January 11, 2014, accessed January 17, 2017, https://www.nytimes.com/2014/01/12/business/in-richmond -california-a-long-shot-against-blight.html?_r=0.

17. Native Times, "Native Groups Partner to Promote Green Careers," January 19, 2009, accessed April 24, 2017, https://www.nativetimes.com/jobs/22-life/education/894-native -groups-partner-to-promote-green-careers.

18. Kentuckians for the Commonwealth, http://www.kftc.org, accessed January 17, 2017.

19. Gopal Dayaneni, quoted from presentation in Whitakers, North Carolina, August 18, 2016.

20. Movement Generation, *Redefining Resilience: Principles, Practices and Pathways,* accessed January 28, 2017, http://pathways-2-resilience.org/ebook/part-ii-redefining-resilience.

CHAPTER 4

Energy Democracy Through Local Energy Equity

STRELA CERVAS AND ANTHONY GIANCATARINO

In 2011, women representing working-class communities of color sought to do something bold—sponsor legislation that would bring renewable energy into environmental justice communities. Working as members of the California Environmental Justice Alliance (CEJA), they sponsored legislation called Solar for All. Solar for All called for a massive transformation of our energy system: away from natural gas power plants and oil extraction and into an energy economy rooted in local, decentralized renewable generation, equity, and community leadership. As the first-ever CEJA-sponsored bill, it put impacted communities at the center of the solution and advocacy efforts. For the first time, women of color walked the halls of the California State Capitol, going toe-to-toe with fossil fuel industry lobbyists and led on the ground, organizing hundreds of community members to advocate for the bill.

Although, the industry killed the bill in 2011, CEJA's efforts transformed the paradigm of how energy and environmental policy is made in California by shifting power from the fossil fuel industry and utilities to communities. In 2015, CEJA partnered with California State Assembly member Susan Eggman, the California Solar Energy Industries Association, and Everyday Energy on a similar bill, titled the Multifamily Affordable Housing Solar Roofs Program (AB 693). The bill, signed into law later that year, built upon the previous goals of Solar for All. It prioritized environmental justice communities to be first in line to get the benefits of solar. The program called for an unprecedented $1 billion investment in solar on affordable housing in low-income neighborhoods, the largest investment in solar energy for disadvantaged communities of color in the country.

This victory shows that a movement, when organized, can take on the biggest lobbyists, can envision equitable energy policies, and can lead the way to create clean renewable energy neighborhoods for all. It is a testament to the power that communities of color can have in creating just solutions and policy ideas that can lead to statewide transformation. Though specific to California, the story elevates critical principles, strategies, tactics, and goals that communities are employing at local and state levels to build political, economic, and literal power—as a solution to climate change, to environmental injustice, and to an extractive energy system.

In this chapter, we will flesh out what this story means in the context of two states that sit at opposite ends of the renewable energy transition: California and Pennsylvania. We will examine a "green zones" policy framework for local energy equity that prioritizes environmental justice and community-of-color leadership and governance. We will identify the barriers that stand in the way of making energy equity possible. And we will share insights from two models of this policy framework for local energy equity taking shape in two vastly different states.

A Policy Framework for Local Energy Equity

The road to energy democracy must, and should, demand progressive, racially and economically just policies at all levels that can be leveraged to take on the corporate power structure, fight extraction, and push against the green divide that is taking shape across the nation. But as a right-wing federal government proposes to expand investments in fossil fuels and privatized energy systems, state and local solutions are a paramount strategy for democratizing energy.

Local policy organizing has the flexibility to experiment with different solutions; allows for community participation and control in the decision-making process; can identify contextual challenges that need to be addressed; and ultimately leads to lessons that can create potential frameworks and guides that can inspire others to adopt change. Building change at the local and state levels can be a catalyst for change nationally, as communities share practices, lessons, and strategies that can generate momentum to transform public policy and investment.

There are multiple efforts advancing a racially, economically, and socially just transition away from an extractive energy economy into a more sustainable and localized economy, rooted in collaboration with labor, community, and strong racial justice principles. Some of these are elaborated in the just transition framework of Movement Generation described in chapter 3 and the work of the Asian Pacific Environmental Network described in chapter 5. In this chapter, we offer a policy framework that creates healthy and thriving communities, called "green zones" (or "energy investment districts"—EIDs, as they are referred to in our Philadelphia example below).

Green zones provide a framework for equitable energy policies by infusing impacted communities with the financial and technical assistance needed for development of local renewable energy resources. Green zones host impacted communities—most often low-income communities and communities of color—in which residents organize to reduce industrial pollution and to cultivate new, coordinated opportunities to implement community-based solutions.

While each green zone is a reflection of the specific needs, priorities, and environmental justice issues of the community, it shares common roots. By identifying green zones—communities that need to transition from industrial pollution into healthy neighborhoods—we can advocate for policies that direct a whole range of resources into programs in those communities. There are five aspects to the green zones/EID equity model:[1]

1. **Identifying overburdened and impacted communities.** Clear criteria and definition are needed to identify where communities of need are in order to target investment to communities of color and low-income communities that have suffered inequitable environmental and economic hardships. In California, the Communities Environmental Health Screening Tool version 2.0 ("CalEnviroScreen 2.0") combines nineteen different indicators of pollution, toxics, and socio-

vulnerability factors to identify communities that are the most polluted and impacted. In Philadelphia, a standard or robust metric has not yet been established; however, the city's Rebuild initiative offers a cumulative impact assessment (that measures socioeconomic data points and health points) to identify communities of impact.

2. **Prioritizing identified communities for public investment.** Once overburdened communities are identified through a governmental metric, the state or city should then develop a comprehensive policy to direct regulatory attention, job creation programs, and economic development initiatives to these communities that puts them first in line for resources to transform their communities into healthy thriving green zones.

3. **Advancing on-the-ground models:** When applying the green zones model, we should prioritize local projects that install local renewable energy directly in overburdened census tracts. Local models should prioritize giving good local jobs with sustainable wages to the locally impacted communities. Local models should transform neighborhoods and ensure sustainable development that does not result in the displacement of longtime residents or businesses.

4. **Providing resources and assistance to impacted communities.** Identify significant pots of resources and technical assistance that will be directed to fund robust community projects to comprehensively transform toxic hotspots into green zones.

5. **Establishing community governance and democratic decision-making processes.** Build and cultivate community governance and decision-making processes to ensure that resources going into a community and revenue generated by projects are determined by community input, leadership, and plans.

Barriers to Achieving Local Energy Equity

The green zones/EID policy framework expressed in the previous section is meant to promote local energy equity through state and city policies that provide targeted resources to impacted communities. A number of barriers are commonly encountered in moving these policy goals in this direction:

1. **Lack of "political will."** The fossil fuel industry and lawmakers have historically discriminated against and disinvested in environmental justice communities when it comes to renewable energy. These communities are not a priority for them, and they leave the clean renewable energy for more affluent white communities. We have seen a huge inequity of renewable energy infrastructure getting sited in wealthier single-family homes rather than in communities that need these investments the most.

2. **"Green divide."** Although some states, like California, experience an increase in renewable energy, much of the infrastructure has been large-scale solar farms and wind turbine farms in the desert. Although large-scale solar farms generally meet our overall renewable energy goals, these farms do not address the lack of renewable investment and benefits in local communities that need it the most. Environmental justice communities experience a "green divide," in which renewable energy is being developed, but the green energy is being built far away from communities that need it the most, leaving these communities again neglected and without the benefits of renewable energy, including local jobs.

3. **Financial costs.** The costs associated with renewable distributed energy systems are a significant barrier for environmental justice communities. Many environmental justice communities cannot afford the large up-front costs of rooftop or community solar and other renewable energy. Installing renewable energy systems includes the high price for the equipment as well as "soft cost," such as permitting and interconnection fees and taxes. Additionally, many low-income customers do not meet the credit requirements to obtain low-cost financing or affordable loans and leases.

4. **Ownership.** The high number of low-income communities that rent homes or apartments is another significant barrier. California has one of the lowest homeownership rates in the country—54.3%.[2] Lower-income households, of which many are found in the most disadvantaged communities, make up almost two-thirds of all renter households and often pay more than half of their incomes toward rents.

In Pennsylvania, homeownership rates have steadily declined for the past twelve years.[3] And in Philadelphia, the rates have declined the most steeply as compared with any major city,[4] especially for black homeowners.[5] The most disadvantaged communities can contain disproportionately high rates of renter-occupied housing units. Additionally, many of these low-income residents live in multi-unit residential rental properties, which limits their ability to approve of and have ownership of the renewable energy. This is especially true when the landlord would be paying for the system, but the tenant would benefit from the reduced energy costs.

5. **Outreach, education, and marketing.** Proper outreach, education, and marketing present a barrier to getting renewable energy in environmental justice communities because of the unique characteristics of these communities. Residents need outreach to be conducted by people they trust—primarily from established community-based organizations that have a history with organizing in the local community. Outreach, education, and marketing must be conducted in the language of the community and must be culturally appropriate. Additionally, organizations must translate materials so that they can truly reach communities in need of the renewable energy. Many community members also do not have internet access. Face-to-face outreach and culturally appropriate and multilingual marketing are best for environmental justice communities.

6. **Resource sustainability.** A major challenge to the success of energy equity initiatives in our current economic and energy system is the resource gap that does not allow communities to participate fully without draining capacity away from other work. The resource gap in communities makes it difficult to take on energy democracy efforts alone as an individual, such as prioritizing leaky roofs, or organizationally, such as resourcing organizers and implementers. Thus the most successful efforts addressing energy equity tend to cross sectors and issues that both build on existing work and challenges and also leverage resources most effectively and sustainably.

Advancing Locally Rooted Energy Equity Initiatives

In this section we examine two examples in which historical conditions have set the stage for advancing local energy equity: one in California and one in Philadelphia, Pennsylvania. In each case, we explore the historical legacy and then point to efforts to overcome the barriers discussed in the last section.

CALIFORNIA: ACHIEVING ENERGY EQUITY IN A PROGRESSIVE POLICY LANDSCAPE

California is important to examine as it has led the nation in advancements in renewable energy and climate justice policies. California has set aggressive renewable energy standards and has explicitly focused on environmental justice communities to direct renewable energy investments in state policies. California is also home to some of the prominent environmental justice organizations that have led the movement to advance these policies.

The California Context

In California, the takeover of land and community is witnessed all across the state as utility companies become megacorporations operating for huge profits at the expense of community health and well-being. Environmental justice communities have led heroic fights to shut down the fossil fuel industry and demand clean, local, renewable energy.

In the city of Oxnard—a community with a majority Latinx[6] population and that, according to CalEnviroScreen, ranks in the top 10% of most polluted and impacted census tracts—the Central Coast Alliance United for a Sustainable Economy (CAUSE) and CEJA have led a campaign against energy companies that have taken over to build dirty natural gas power plants in the area (figure 4-1). Energy companies such as NRG and Southern California Edison (SCE) have pursued a massive build-out of unnecessary gas power plants in an already overburdened community.[7] Oxnard is home to three power plants, three landfills, and a Superfund site where a metal recycling plant left a wake of toxic chemicals. NRG and SCE claim that they need these power plants to fulfill the energy need in the area and to replace the phaseout of nuclear power plants in California. However, the energy need in California can be met with renewable energy instead of dirty natural gas.

The historical practice of energy companies siting power plants in low-income communities of color is the ultimate form of environmental racism, and CAUSE has mobilized hundreds of community members to topple this system, demanding to replace these power plants with renewable energy. These power plants are known to release dangerous amounts of methane and co-pollutants into the local community, endangering public health, causing climate chaos, and perpetuating the legacy of neglect for the community while benefiting the utility corporations.

In Southeast Los Angeles, Communities for a Better Environment fought back against an industry that attempted to build a massive power plant that would have released 1.7 million pounds of chemical emissions and soot particles into the surrounding Latinx community.[8]

In Otay Mesa, a heavily low-income Latinx community near the Mexican border in San Diego, the Environmental Health Coalition led a campaign against San Diego Gas & Electric, which had pushed through an unnecessary $1.6 billion gas power plant that will release an estimated 40 million tons of carbon emissions.[9]

Richmond, California, is home to the Chevron oil refinery, and Wilmington, California, is home to the Tesoro and Valero oil refineries. These oil refineries are sited in majority Latinx, Laotian, and black neighborhoods whose residents live below the poverty level. These refineries are notorious for hazardous environmental pollution and health impacts. Their disastrous explosions have sent numerous community members to the hospital. Asian Pacific Environmental Network and Communities for a Better Environment are leading a long-term campaign to create local energy solutions that can benefit the community and prevent the destruction of the air and land, challenging refineries that enjoy windfall profits at the expense of community health.

In the city of Shafter in California's San Joaquin Valley, again a heavily Latinx community that ranks in the top 20% most polluted zip codes,[10] big oil is taking over the area with the quick build-out of hydraulic frac-

Figure 4-1. Members of the Central Coast Alliance for a Sustainable Economy (CAUSE) protest in front of the California Public Utilities Commission against licensing of a dirty natural gas power plant. The power plant is the fourth plant expected to be built in the city of Oxnard. *Image source: California Environmental Justice Alliance*

turing (fracking) operations. These fracking wells surround schools, churches, farms, and homes. The dangerous chemicals used in fracking contaminate the soil, water, and air, resulting in headaches, rare cancers, and respiratory diseases and rendering the water undrinkable. The Center on Race, Poverty & the Environment is leading a grassroots campaign against these fracking operations and calling for justice in climate and energy solutions that will benefit the local community.

The California Green Zones Initiative

A unique blend of strong community-based resilience, organizing and leadership with strong elected leaders, and a history of environmental activism make California a national leader and model for other states.

The California Environmental Justice Alliance (CEJA) is the foremost multiracial environmental coalition in California, representing eleven different organizations. CEJA has taken California by storm, advancing visionary grassroots-led policies and helping to pass dozens of environmental, energy, and climate policies that put equity at the center.

The Green Zones initiative was launched by CEJA after years of attempting to organize against individual sources of pollution and toxic sources. CEJA members realized that rather than trying to fight off these individual polluters, what was needed was a comprehensive approach to addressing environmental inequities while driving resources into highly impacted communities rather than a "whack-a-mole" approach. The green zones approach allows community residents to envision how their communities can be transformed into clean sustainable communities that encompass local renewable energy, affordable housing, community gardens, clean air, and clean water.

To make green zones possible, California policymakers have adopted a "California Communities Environmental Health Screening Tool" ("CalEnviroScreen"). Developed by the Office of Environmental Health Hazard Assessment under the California Environmental Protection Agency, CalEnviroScreen is a comprehensive cumulative impacts tool that can identify some of the most overburdened environmental justice communities in California.

The development of the CalEnviroScreen and the Green Zones initiative greatly influenced the Multifamily Affordable Housing Solar Roofs Program (AB 693) discussed in the chapter opening. This program includes the following components:

1. **Creating a robust program, not just a pilot program.** Historically, California lawmakers advanced "equity" by carving out small funding streams for impacted communities to participate in renewable energy. However, this approach is not enough to achieve true

energy equity. With AB 693, CEJA pushed for a bolder development of an intentional and robust program that would direct significant funding toward these impacted communities. The Multifamily Solar Roofs Program[11] ultimately created a 300-megawatt renewable energy program specifically for disadvantaged and low-income tenants of multifamily affordable housing, the largest of its kind in the nation. This ten-year program is expected to benefit 150,000 low-income renters.[12]

2. **Utilizing a scientific methodology to identify the most "disadvantaged" communities.** The Multifamily Solar Roofs Program includes the CalEnviroScreen definition of a "disadvantaged community," in which projects will be targeted in the top 25% of the most impacted census tracts. Getting the definition of a "disadvantaged community" is critically important to move the needle of renewable energy programs past the current 6% penetration in disadvantaged communities.

3. **Infusing significant funds into environmental justice communities.** The program taps into the electricity sector cap-and-trade revenues, a source of funding that is meant to bring renewable energy to low-income communities but is currently not serving its function. The program will invest $1 billion over ten years for renewable energy on multifamily affordable housing, the largest investment of its kind for disadvantaged communities.[13] This policy also uses the green zones initiative goal of identifying a large pot of funding for the program. It also addresses the huge financial barriers within low-income communities to accessing renewable energy.

4. **Focusing the economic and jobs benefits onto impacted residents.** A significant proportion of environmental justice communities in California are renters. These renters face a myriad of barriers to accessing and benefiting from renewable energy programs. Economic and jobs benefits are some of the significant barriers for renters where the current renewable energy programs for multifamily housing incentivize the landlord but have no economic benefits for the tenant. The Multifamily Solar Roofs Program utilizes a "split incentive" in which both

renters and landlords will get the economic incentive of a reduction on their bills, thus attracting both types of populations to the program. Moreover, the program includes a local-hire mandate to provide economic development benefits to communities that are suffering the highest rates of unemployment.

5. **Building local, decentralized generation.** To address "green divide" barriers, in which large-scale renewable energy facilities are often built outside of impacted communities, this program is designed to localize renewable energy projects by installing small-scale local projects in the most highly impacted census tracts.

6. **Transitioning away from fossil fuels.** The Multifamily Solar Roofs Program will help California transition away from fossil fuels and help meet the state's new goal of achieving 50% renewable energy by 2030. The program will significantly reduce greenhouse gas emissions and the need for dirty natural gas power plants.[14]

PHILADELPHIA: ENERGY EQUITY IN A CHALLENGING POLICY LANDSCAPE

Unlike California, where the policy and political context is much stronger for energy democracy initiatives, the context in Philadelphia is behind the curve. However, there are creative efforts under way that can lead to policy change around energy, financing, and land use through an energy investment district model that shares the green zones framework and principles.

Pennsylvania is a textbook example of a state with an extractive energy economy. Towns were built on the mining, processing, and shipping of coal. Today, fracking and shipping oil and gas by rail drive the energy economy. The extraction has left its toll on communities and has greatly hindered the development of strong state policy in this area.

The state lacks significant criteria and data collection to identify impacted communities in the way California's EnviroScreen tool allows. The state's transition to renewables is inadequate, requiring only 18% alternative energy by 2021, and includes fuel sources, such as municipal converting solid waste to energy, that still have negative impacts on environmental justice communities. The policies fail to create economic democracy and wealth building. For example, the Pennsylvania Public Utility Commission's proposal for a 200% net metering cap that would have allowed residents to create a revenue stream from excess generation was overturned by the state's Independent Regulatory Review Commission. Virtual net metering and community solar projects are not allowed for multiple owners of meters. Lastly, there are no real financial incentives to capitalize a project or technical assistance support for project development, leaving people to rely solely on the federal tax credit program.

The Philadelphia Context

The city of Philadelphia is a key focal point for piloting on-the-ground models to create energy equity and shift state policy. It is the largest city in the state; it does not have a shortage of "disadvantaged communities" as discussed in the green zones framework; and the city's community-based organizations are building power to scale out a statewide effort for economic justice in energy.

With a staggering poverty rate of 26.3%, Philadelphia is known as the largest poorest city in the nation, and the majority of residents that live below $10,000 are black.[15] Philadelphia also has one of the highest incarceration rates in the nation, and its asthma rate is almost twice the national average.[16] It is no coincidence that the city is also home to the largest oil refinery on the East Coast, the Philadelphia Energy Solution (PES) refinery, located in Southwest Philadelphia.[17] Lead hazards, gas leaks, and in-home environmental hazards are highest among lower-income populations of color in north, west, and northwest Philadelphia.[18]

One statistic can sum up the impact of racial inequity within the city: life expectancy. In the Old City neighborhood, where the population's median income is the highest in the city at over $100,000 and is 90% white,[19] life expectancy is eighty-eight years of age.[20] Just west of Old City, in Parkside, a neighborhood that is 90% black and has an average median income of $15,000,[21] life expectancy is thirteen years less. One in two people faces energy and food insecurity.[22] The neighborhood is a microcosm of the green divide and has ownership barriers to participation. In addition to the policy challenges identified earlier, the relatively low homeownership rates and the older housing stock as compared with the rest of the city limit neighborhood residents' ability to participate in the renewable energy economy. Among both owners and renters in the neighborhood, nearly 53% of homes were built prior to 1939, and nearly 70% of all homes were built before 1960.[23] As one might expect with older housing stock, structural maintenance, such as roof repairs, is prevalent and takes priority over solar installations. Further, the lack of investment and policy supports makes it difficult for homeowners with little disposable income to even access the solar economy.

Energy Investment Districts: A Green Zones Approach in Philadelphia

Despite the challenges, Parkside is rich in its vision, local organizing, and work and commitment to equity. Residents are working to develop an energy investment district as a way to build wealth and create opportunity that takes on the economic reality and can help shift policy for other impacted communities to lead in the energy transition. The community has assets to lead in the renewable energy transition. There are vacant properties and community spaces that could be prime for a community solar farm generating a potential of 5 megawatts of solar. Such a project would produce revenues of $700,000 annually for a community-based nonprofit in which the governance and deployment of the generated revenue is directed by neighborhood residents for its residents.[24]

And this model is important. It is a fundamental shift in not just who benefits from solar, but who owns and controls the economy of solar. It actively takes on the barriers of a green divide, financing, ownership, and resource sustainability and challenges the political will of the status quo. To the Parkside community, leaders see renewables as a strategy to facilitate true "equity" ownership (monetary, psychological, emotional, cultural, and social). The energy investment district (EID) is about restructuring the energy economy by prioritizing and valuing *how* projects are developed, *who* develops them, and *how* they are financed.

The Parkside EID follows the green zones framework, particularly by identifying an impacted community that can advance on the ground models through resourcing community leadership and establishing community governance. While there is no formal city or state program to identify communities like the CalEnviroScreen, Philadelphia's Rebuild initiative[25] has aggregated economic, race, and social data to identify communities for investment. Parkside is one neighborhood that stands out for deeper city focus.

The Parkside EID aims to resource and invest in community governance and ownership. While current policies do not allow for true community-owned solar projects, the Centennial Parkside Community Development Corporation (CDC) solution is to be the direct owner and operator of solar systems on strategically selected vacant land within the neighborhood's boundaries and to enter into long-term power purchasing agreements with anchor institutions to attract sustained revenues, public investment, foundation investment, and social capital to make the plan a reality. While the CDC would be the owner in name, it espouses the green zones framework by placing the planning, decision making, and resource allocation of the project squarely within the community's hands. One approach to achieving this goal is to run a participatory budgeting process that allows residents to decide how revenues are utilized and how the project could expand over time. Like communities in California's green zones, residents in the neighborhood want to see an increase of resources in order to tackle energy efficiency and home repairs; address energy and food insecurity; invest in place-based improvements; and, most important, generate wealth and economic development so the neighborhood does not get displaced by the creeping gentrification that is occurring around its boundaries.

How the EID Aims to Tackle Barriers to Energy Equity

While the Parkside Energy Investment District model may be limited in its impact in the short term, as a creative model it offers the opportunity of shifting policy that can lead to long-term impact. Good policy often follows good models, and Philadelphia can provide such a model. Models such as the Parkside EID can transform some of the barriers to energy equity though policies that specifically tackle:

1. **Financing.** The biggest challenge for the Parkside EID is financing for implementation. Part of the implementation process can expose the shortcomings of the state's current policy and the lack of financial investment in a local energy solutions. This model can surface the challenges of financing renewables in a state where electricity rates are low and highlight the shortcomings of the existing renewable energy credit and current net metering standards.

2. **Political Will**

   a. **Municipal procurement policies across the state.** Municipalities can fill the gap of state governments through procurement and leadership. For example, the City of Philadelphia spends about $60 million annually to procure electricity that leaves the city to outside power providers and companies.[26] Most procurement policies must act within fiduciary responsibility and often lead to spending the least amount of money for the most product in terms of electricity. The Parkside EID model makes the case that renewable energy production should not just be measured in cost per kilowatt-hour but should take into account the greater impact on the city as part of the value and cost of procurement. This could include community wealth building; local job creation; health and well-being of the city's most vulnerable populations; environmental protection; efficiency; racially and economically just and equitable labor practices and economic development.

   b. **Strong renewable procurement standards.** The EID can make the case for a much stronger renewable procurement standard (RPS) that not only prioritizes renewable energy development but calls out specific standards around prioritizing low-income communities, environmental justice impacts, and local ownership of solar. Such models can be found in Washington, DC's most recent RPS, which requires that the district be powered by 50% renewables by 2032, with a specific focus on prioritizing and investing in the 100,000 low-income households.[27]

3. **Ownership.** The EID challenges the concept of ownership and could seed a program similar to California's Multifamily Solar Roofs Pro-

gram. While California is much further along than Pennsylvania, the elements of the Solar for All legislation are being played out in this organizing approach by Parkside. This model can elevate the systemic challenges in Pennsylvania and show how communities are thinking creatively to provide new solutions for how impacted residents can participate as owners in the energy economy.

Conclusion[28]

An energy system that is democratic and equitable is possible. We have an opportunity to shift away from the archaic profit-driven fossil fuel economy model to one that is envisioned by the community and owned and operated by the people. Our energy system has the potential to benefit the overall public health and local economy and can create long-term sustainable jobs.

In order to achieve this vision of transitioning from fossil fuels to local, democratic renewable energy, we must insist on organizing as a centerpiece in this endeavor. As the United States becomes a majority people-of-color nation, it will be all the more imperative to address the plight of those that are hit hardest by climate change and dirty energy. It will be people of color who will envision an alternative to fossil fuels and people of color who will lead this transition to our clean energy future. California and Pennsylvania provide two examples of why people of color must build on the progress we have made, continue to fight back against industry and environmental racism, and lead the movement toward a more equitable and just energy system. We need to find ways to build the energy transition that roots and centers people of color and can create strong enough local and state policy solutions that can lead the way in a time of federal retrenchment.

We need to identify what policies can create levers for greater change. In this vein, the movement for energy democracy must hold an intersectional lens that lets us see how our energy system plays within a greater system of issues. This lens connects race, class, and gender analyses to equitable energy policies and programs. In doing so, this lens will create different ways of thinking about energy solutions and ultimately a stronger community-owned process and outcome. Energy policies must truly benefit all people, and to achieve this end, these policies must target communities that have historically been locked out of the benefits of our energy system. Therefore, energy policies that get us to energy democracy can draw from the intersectional approaches of California's green zones initiative and Philadelphia's energy investment district to develop equitable energy principles and address long-standing barriers to accessing renewable energy. If communities of color continue to be excluded from energy policies, the United States has no chance of transitioning away from the exploitative fossil fuel system or of achieving our national and state climate and energy goals.

1. Kay Cuajunco and Amy Vanderwarker, "Green Zones in California: Transforming Toxic Hotspots into Healthy Hoods" (California Environmental Justice Alliance, 2015).
2. Federal Research Bank of St. Louis, *Homeownership Rate in California,* accessed January 12, 2017, https://fred.stlouisfed.org/series/CAHOWN.
3. Federal Research Bank of St. Louis. *Homeownership Rate in Pennsylvania,* accessed January 12, 2017, https://fred.stlouisfed.org/series/PAHOWN.
4. Pew Charitable Trusts, "State of the City: Philadelphia" (Pew Charitable Trusts, 2015), accessed January 12, 2017, http://www.pewtrusts.org/~/media/assets/2015/05/2015-state-of-the-city-report_web.pdf.
5. Pew Charitable Trusts, "Homeownership in Philadelphia on the Decline" (Pew Charitable Trusts, 2014), accessed January 12, 2017, http://www.pewtrusts.org/~/media/assets/2014/07/pri-homeownership-report_final.pdf.
6. We use the term *Latinx* rather than *Latino/a* to be inclusive of people who may identify as transgender and gender nonconforming.
7. Lucas Zucker, "Dirty Power's Last Stand?," *Dissent* (2016), accessed January 12, 2017, https://www.dissentmagazine.org/online_articles/oxnard-california-nrg-last-fossil-fuel-power-plant.
8. Communities for a Better Environment, accessed January 12, 2017, http://www.cbecal.org/about/victories/.
9. Kimberly Tomichich, "Proposed Pio Pico Power Plant: A Costly 25-Year Mistake San Diegans Cannot Afford," *San Diego 350* (2013), accessed January 12, 2017, http://sandiego350.org/blog/2013/12/11/proposed-pio-pico-power-plant-costly-25-year-mistake-san-diegans-afford/.
10. Author analysis of data from Office of Environmental Health Hazard Assessment, "California Environmental Screen" (2016), accessed January 12, 2017, *http://oehha.ca.gov/calenviroscreen*.
11. Strela Cervas, "Is Your Building Eligible for the New Multifamily Solar Roofs Program?," *Medium* (2016), accessed January 12, 2017, https://medium.com/@cejapower/is-your-building-eligible-for-the-new-multifamily-solar-roofs-program-21e508643ce3#.j0pzk1scw.
12. Raven Rakia, "Solar Power Access Looking a Lot Brighter in California," *Grist* (2016),

accessed January 12, 2017, http://grist.org/climate-energy/solar-power-access-looking-a-lot
-brighter-in-california/.

13. Kat Friedrich, "California Supports Solar Roofs for Multifamily Affordable Housing," *Clean Energy Finance Forum* (2016), accessed January 12, 2017, http://www.cleanenergyfinanceforum
.com/2016/07/18/california-supports-solar-roofs-for-multifamily-affordable-housing.

14. Cervas, "Is Your Building Eligible for the New Multifamily Solar Roofs Program?"

15. Philadelphians Organized to Witness, Empower and Rebuild (POWER), "Black Work Matters: Race, Poverty and the Future of Work in Philadelphia" (2016), accessed January 12, 2017, http://powerinterfaith.org/wp-content/uploads/2016/03/Black-Work-Matters
-Report.pdf.

16. Drexel University, "Five Things to Know About the Role of 'Place' in Asthma Research" (2015), accessed January 12, 2017, http://drexel.edu/coas/news-events/news/2015/February
/5-things-about-role-place-asthma-research/.

17. Philadelphia Energy Solutions, accessed January 12, 2017, http://pes-companies.com
/refining-complex/history/.

18. Barbara Laker, Dylan Purcell, and Wendy Ruderman, "Philly's Shame: City Ignores Thousands of Poisoned Kids," *Philadelphia Inquirer/Daily News* (2016), accessed October 16, 2016, http://www.philly.com/philly/news/Philadelphia_ignores_thousands_of_kids_poisoned
_by_lead_paint.html.

19. Author analysis of U.S. Census Bureau, "American Community Survey 2015: 5 Year Estimates" (2016), accessed October 16, 2016.

20. Virginia Commonwealth University, Center on Society and Health, "Philadelphia Life Expectancy Methodology and Data Table" (2016), accessed January 12, 2017, http://societyhealth.vcu.edu/media/society-health/pdf/LE-Map-Philly-Methods.pdf

21. Author analysis of U.S. Census Bureau, "American Community Survey 2015."

22. Virginia Commonwealth University, Center on Society and Health, "Philadelphia Life Expectancy."

23. Author analysis of U.S. Census Bureau, "American Community Survey 2015."

24. Estimates provided by the Centennial Parkside CDC.

25. The Rebuild initiative is a multiyear billion-dollar effort to invest in public infrastructure of parks, schools, and public spaces. While it does not tackle issues like energy, Philadelphia collected various data and analysis to do an equity score to identify priority areas for investment. For more information, see http://rebuild.phila.gov/about/learn-about-the-data #Equity Factors.

26. City of Philadelphia, Office of Sustainability, "Energy Benchmarking Reports" (2016), accessed January 12, 2017, https://beta.phila.gov/documents/energy-benchmarking
-reports/.

27. DC City Council member Cheh, "Renewable Portfolio Standard Expansion Amendment Act of 2016" (2016), accessed January 12, 2017, http://lims.dccouncil.us/Legislation
/B21-0650.

28. "The California Energy Commission released a Barriers Report that comprehensively outlines the barriers and opportunities to accessing renewable energy and energy efficiency for disadvantaged communities" available at http://www.energy.ca.gov/sb350/barriers_report/

Base-Building and Leadership Development for Energy Democracy: APEN's Work in East Bay Asian Immigrant and Refugee Communities

VIVIAN YI HUANG AND MIYA YOSHITANI

The reason I'm here to support SB 350 is because it will reduce fossil fuel emissions by 2030, and this is important. I used to study chemistry, so I know that the emissions and pollution from the refineries can be poisonous for people. There are chemicals that can cause blindness, and pollution can really harm our bodies. Reducing fossil fuels is really important for improving our health, so this is for all of us to have good health. This is why I'm very much in support of SB 350.

APEN member Lisa Cheng[1]

We speak for ourselves. This is a defining principle of environmental justice, and a core value in centering frontline leadership at the forefront of an energy and economic transformation. When we speak for ourselves, energy democracy is not just a change from dirty energy to clean, renewable energy, but a refocusing on the needs, solutions, and leadership of real people, especially those most impacted by the worst of our energy system, and a commitment to self-determination as a moral and a strategic imperative.

When energy democracy emerges from the lived experiences of community residents and is grounded in a long history of organizing for solutions at the intersection of racism, poverty, and pollution, it goes well beyond mainstream efforts to incentivize, finance, and build more renewable energy and demonstrates a pathway for transforming our energy economy.

With her testimony, Lisa Cheng demonstrates the power of organizing and leadership development, showing how community voice and power are essential to transforming our current energy economy. Throughout history, the use of this energy has been most decidedly undemocratic, exploitative, and extractive. As one example, Chevron holds an annual shareholder meeting, a meeting comprised primarily of investment fund managers and corporate executives—people who are not living on the front lines of the corporation's pollution impact. The shareholders get an update on the company's profit margins, hear about the company's future direction, and vote on resolutions. The focus of their decisions is not on the health of the planet, the land, or the surrounding communities, but rather on investment returns, the profits to be gained, and the company's projected quarterly earnings. This corporate process is what our economy currently has in place for "the energy economy."

The current energy economy does not work for people or the planet, given the social, health, and environmental costs of the fossil fuel industry. What we actually need is a radically different political, economic, and environmental platform that is fundamentally grounded in true democracy. We need to reshape our relationship to energy and economy, and to each other and the planet, a transition that will not happen without building a real base of people power and the development of people's leadership.

Engaging Asian Americans in Energy Democracy

The Asian Pacific Environmental Network (APEN) was formed in 1993, following the First National People of Color Environmental Leadership Summit, to organize and build the power and voice of low-income immigrant and refugee Asian American communities in the San Francisco Bay Area for environmental justice. Like the hundreds of community-based environmental justice organizations around the country, APEN's base-building work has been the heart of the organization and is the foundation for our successful campaign wins. From the state capitol to the ballot box to the community center, APEN members have spent decades creating a vision of energy democracy, advancing alternatives, and expanding the movement of people organized around this agenda.

Our base-building work has been rooted in two different communities with multiple Asian ethnic groups speaking different dialects and languages. We have been honored to work with the Laotian refugee community in Richmond, California, and the Chinese immigrant community in Oakland, California.

RICHMOND, CALIFORNIA

Richmond is home to a diversity of cultures and communities, including thousands of Laotian refugees. The city is also home to the Chevron oil refinery, one of the largest greenhouse gas emitters in the state. In the shadow of this refinery, 80% of community residents are working-class people of color, and 25% live below the federal poverty line. The refinery's pollution has created deep and major health impacts, and the voices of community members show the toll of being on the front lines.

Richmond resident Koy Seng Saechao remembers the refinery's explosion in the 1990s: "I experienced strong headaches and dizziness. Sometimes, I still experience headaches and irritation of my eyes. It is hard for me to breathe and my body aches."[2]

Another longtime Richmond resident, Lena Phan, says: "I've been in Richmond pretty much all my life. I was one of the many children who grew up in North Richmond. With its low-income housing, it was our only option at the time. Now, I work just blocks from the Chevron refinery. As I watch the fumes roll out of Chevron's smokestacks, I vaguely remember having to 'shelter in place' as a child during fires, explosions, and other emergencies, closing our doors and windows, hoping not to breathe in the toxics. It always seems to be our neighborhoods, low-income communities of color, that suffer when big corporations want to build, expand, and pollute our cities."[3]

Despite the tough health and economic conditions faced by the refugee community, their strength and resilience have been formidable in standing up for environmental justice.

CHINATOWN, OAKLAND, CALIFORNIA

Oakland's Chinatown is a vibrant community full of culture, despite a history of continual acts of displacement. The first Chinese immigrants, who formed various Chinatowns in Oakland during the 1850s, had been driven from the fields of the Gold Rush by racist miners. Later, in light of intense racism, some of these Chinatowns would be forced to move or eliminated altogether.

Because of its central location near downtown, Oakland's Chinatown has continued to face threats over the years. In 1965, three blocks in Chinatown were demolished, and Madison Park was relocated to create the Lake Merritt Bay Area Rapid Transit (BART) station and BART headquarters. Destruction in this process included the homes of seventy-five families, a former orphanage for girls, and the Chinese True Sunshine Episcopal Church.

In more recent years, the neighborhood has faced threats from planning and development that does not prioritize affordable housing and tenant protections. Community members highlighted the continued community challenges and their hopes for a better future.[4]

Li Ya Chen, a member of APEN, says: "When I first arrived to the Bay Area, my monthly income was only about one thousand dollars, but the rental of a one-bedroom apartment was over seven hundred dollars. It was very difficult to survive after paying the rental. I heard about low-income housing, but I was helpless and at a loss because I didn't know where and how to apply."

As Shao Yang Zhang, a member of APEN, sees it: "I believe public safety and job and housing issues are related to each other. In order to solve the problem on public safety, we must solve the job and housing issues as well, because they are the root causes of the problem. The more housing complexes, the more supporting facilities you will have to provide more employment opportunities to the local community and help enhance the people's standard of living."

Although many immigrants in Chinatown are monolingual, they have lifted up their voices repeatedly to champion climate resilience and clean energy jobs to support an environment and local economy that allows people to stay in their homes and the neighborhood to thrive.

Building Power: Base Building and Leadership Development

Ask any organizer, and they will have a story of how they came to be involved in organizing work. For Vivian, it was becoming conscious of the issues after the passage of Proposition 187, a racist and devastating

California ballot measure that would have banned social services, health care, and education for undocumented immigrants. For Miya, it was being inspired to become an organizer after witnessing the power of Chicago public housing residents battling industrial polluters surrounding their community and winning.

For Lipo Chanthanasak, an APEN member in Richmond, it was an APEN meeting: "I'm from Laos and arrived to the United States in 1991. I got involved with APEN when I first came to the meeting. After I had come to many meetings, I understood that APEN supports grassroots people to fight for themselves."

What's important about each story is that it highlights the moment that people recognized their power in being part of making change. Organizing aims to foster that recognition of power among many individuals, building collective power for justice.

Unfortunately, too often the discussion about the energy transformation needed to solve the climate crisis focuses on technical solutions and doesn't include an analysis of power. We will not have the political power needed to implement true solutions to climate change, at the breadth and scale they are needed, without engaging millions of people in that fight. And the only way to inspire millions of people to support our solutions is to ensure that we are organizing, block by block, around equitable solutions that reflect the actual needs and aspirations of people at the core. Needs and aspirations include clean air, jobs, generating wealth for local communities, and democratic control over local resources. Transforming our energy infrastructure is as much a project to shift hearts and minds as it is to get solar panels on buildings.

BASE BUILDING

There are as many organizing models as there are communities. Even among the different constituencies within APEN, there are different ways to recruit, retain, inspire, and mobilize members. However, no matter what age, language, or ethnicity, what is important is creating the opportunities and structure for people to engage in democratic decision making.

We have created many different spaces to accommodate people's level of depth and engagement with APEN's work. Members are those who support APEN's work, attend our annual member meetings, and engage in specific ways. They are able to provide feedback to APEN's organizational work and cast a vote in the election for Leadership Steering Committee members.

Those members who are more active—participating in leader meetings, action teams, and campaign events—are called leaders. They are part of decision making in all these spaces, such as developing campaign demands, taking positions on ballot measures, and deciding organizational priorities. For example, in 2011, our organization conducted a process in which leaders learned about different environmental justice issues, discussed them, and then voted on the one that they wanted for the future direction of APEN. In choosing renewable energy, leaders highlighted the opportunity to generate solutions to climate change, to clean the air in their neighborhoods, to create jobs for their communities, and to save money on their energy bills.

At the most engaged level, APEN has a Leadership Steering Committee, comprised of members from both Richmond and Oakland who are elected and nominated by their peers. The Leadership Steering Committee plays a critical role in providing direction and feedback around our statewide work, implementation of organizational priorities, membership and organizing structure, and overall organizing initiatives. As the primary committee that joins both Richmond and Oakland members, they play a critical role in analyzing different state policies and deciding legislative priorities for organizing. They also play leadership roles in their respective local communities (figure 5-1).

LEADERSHIP DEVELOPMENT

Once people recognize they must act to address injustice, there are still many systemic barriers that prevent them from reaching their full leadership potential. Organizing is not simply about mobilizing people for a one-time campaign, but about having a commitment to their development for the long term. Transformational organizing recognizes that we

Figure 5-1. APEN Leadership Steering Committee member Hai Bo Pan testifies to urge the Oakland City Council to ban coal exports from the city. Council members unanimously passed a resolution to do so. *Image source: Brooke Anderson*

need a leader-full movement, that leadership can be developed in everyone, and that people come to the work as both teachers and learners.

In thinking about our own personal leadership development, it's clear that both of us are only who we are today and who we will be in the future because of the opportunities, experiences, and resources that we are continually gaining in the process of developing our leadership. Becoming a leader is not an on-off switch but rather a journey over time through a combination of political consciousness raising, lessons learned from experience, and a desire to make societal change. Just as systemic social change takes time, our collective growth as leaders also takes time.

The Asian immigrant and refugee community members we work with are on a similar journey. Many of them come from a lived experience of injustice—facing physical or economic violence in their homelands, the hardship and racism of being immigrants in the United

States, and the pollution and exploitation rendered by corporations in their neighborhoods. As part of a base-building organization, they can channel those experiences of injustice into actions that push for a better future.

APEN invests in individuals as part of an intentional political program that raises critical consciousness and promotes collective action. The content of the program includes deepening values and political alignment as well as knowledge and skills. We have created different types of trainings, experiences, and activities to support people's development. Our APEN Academy is an eight-session course that engages participants in conversations and skills building around topics such as energy democracy, people's immigration experiences, electoral organizing, and doing outreach. We have also held an APEN Academy 201 version for leaders with more experience to engage more deeply around topics like solidarity among different Asian American ethnic groups, antiblack racism, spokesperson skills, and the just transition framework for building a regenerative economy (figure 5-2). These trainings involve interpretation into Cantonese, Mandarin, Khmu, and Mien. To facilitate deeper learning, there are also APEN academies conducted completely in-language and academies for youth.

Of course, a training by itself is not enough. People's leadership development is continually reinforced through the ongoing activities of our organizing work. Political education and discussion of campaign strategy happens at every leader meeting, and smaller action teams give people the opportunity to practice facilitation and decision making. When there are important current event topics such as comprehensive immigration reform, Black Lives Matter, the Hong Kong umbrella movement, or repression of environmental justice activists abroad, we hold political education workshops and exchanges with activists. This lets our members learn more about the issues and discuss what is happening.

Also, our campaigns offer many opportunities for people to put into practice what they've learned. Whether it's testifying at city council hearings, leading chants at a march, facilitating group discussions on campaign priorities, or speaking at rallies, APEN members embody the

Figure 5-2. APEN Leadership Steering Committee member Lisa Cheng, who provided chapter 5's opening testimony on SB 350, contributes her thoughts to an APEN Academy discussion. *Image source: Brooke Anderson*

leadership that frontline communities are providing to our climate justice movement (figure 5-3).

An important step that is often overlooked is making the time for reflection and support. This year, we have created a member assessment system as a way to better document and guide our members' leadership development growth and goals. On a basic level, the assessment starts with looking at how the members' interests and values align with APEN's strategies, Asian American movement building, and environmental justice. The assessment then moves into looking at the members' knowledge and understanding of our campaigns and key issues. Finally, the assessment looks at the members' skills and experience in areas such as transformative organizing, public speaking, teamwork, facilitation, and training others. We envision this process to be a conversation with each member so that they can provide their own self-assessment of their values, knowledge, and skills along with the organizer's perspective. In this

Figure 5-3. APEN members Zhi Long Zhang (left) and Ming Fong Kyu (center) participate in a Black Lives Matter march. *Image source: Ryan Sin*

conversation, we also want members to express their priorities for leadership development so that the assessment becomes a tool for creating a leadership development plan for each member. As we implement this tool, it will be very helpful to get a sense of the growth of our members, individually or collectively, and in which areas.

Finally, and most important, the fostering of relationships is a foundational component of leadership development. Leadership is essentially about the ability to inspire others toward a collective purpose. Relationships and trust building are key to that inspiration and persuasion. We have fostered many spaces for our members to interact, connect, and build more deeply across geographies, languages, and ethnicities. We hold an annual Leaders Retreat with members from across the organization in which people can learn about the latest analysis of the social justice landscape, provide feedback and decisions around the direction of our work, and spend time getting to know one another. We also host joint exchanges, workshops, and contingents to actions so that members can continue to build with one another. At every Leadership Steering Committee meeting, we have time for a few members to do a "show and tell" in which they bring an object and talk about it as a way to share more about their lives and homelands. We've had members bring in their traditional dress, show photos of their homelands and families, and share their political button collections. What has also been beautiful to see is the connection that members are making across languages, with members teaching each other how to say "hello" and "thank you" in the different languages.

SHARING OUR SUCCESS

Organizing and leadership development work is critical to advancing energy democracy. Without a committed movement of people to this cause, the energy transition will continue to perpetuate the exclusion of frontline community members. The base-building and leadership development work reflects the importance of campaigns being led by frontline community members and the importance of the growth and transformation of people as part of that process.

As Lipo Chanthanasak related earlier in this chapter, many of our Richmond members came to the organization to build grassroots power over two decades ago. Having to go up against a multinational, multibillion-dollar corporation, APEN's members have committed countless hours of strategy, advocacy, and organizing for concrete wins over the past decades. Lipo and his wife, Saeng, remain with the organization because of the wins they have achieved and the community they have built (figure 5-4). After an explosion at the Chevron refinery in 1999, community members organized to create a multilingual emergency warning system. Over a decade later, community members continue to organize to counter proposed expansions of the refinery. Currently, (2017) members are involved in fighting for a region-wide Bay Area Air Quality District rule that would force reductions in refinery emissions and pollution.

Not only do local policymakers respond to the powerful leadership of community members; state policymakers have also recognized the importance of responding to community members who understand environmental justice policy and who have a personal stake in the policy impacts. (See the testimony of Lisa Cheng on SB 350 that leads off this chapter.)

In the movement for energy democracy, we also have to remember that it's not only about the policy win. Energy democracy also requires us to rethink how we relate to one another and how we offer new models of leadership where we are creating power "with" rather than "over" people. Communities don't want to simply engage, they want to be able to govern, to make decisions, and to have agency. Rather than perpetuating the systems of oppression that are in place, we often reflect on how we dismantle those systems in our practice as well.

In supporting and prioritizing the development of community members, our work is recognizing and valuing their wisdom, their beauty, and their capacity for being powerful. Such recognition is part of disman-

Figure 5-4. APEN community organizer Torm Nompraseurt (right) strategizes with Lipo Chanthanasak and his wife, Saeng, two longtime leaders in the fight for renewable energy in Richmond. *Image source: Beth Buglione*

tling systems of oppression that often dehumanize immigrants; refugees; monolingual, illiterate, and working-class community members. When we first conducted the APEN Academy, our Mien and Khmu members, especially the women, felt highly valued in going through that process. Many of them had never been able to have any formal education or learn how to read or write. One member told us how much she appreciated the APEN Academy because it was like school, and she had never had the opportunity to go to school.

It is because of the leadership of our members that our work at APEN has stayed grounded in what people want for their lives and neighborhoods. It was our members who talked about the desire to create tangible models that they could show to their neighbors and families. They pushed us to translate energy democracy into something that could be seen or felt, and pushed our movement to make it real. They identified sites in their communities where they want to see the development of community solar that would bring locally generated energy ownership to more residents. It was also our members who consistently talked about the need to organize more young people into our movement. They were on the forefront of recognizing that it is the young people who are inheriting the climate crises and should be actively engaged in shaping the solutions.

We are constantly inspired by the leadership of our members, and it's heartening to see them get recognized externally as well. In 2013, the White House honored the inspiration of our Richmond leader Lipo Chanthanasak (see figure 5-4) by naming him a White House Champion of Change for Climate Resilience.[5] He was the only monolingual community member to be selected in this cohort. In a video celebrating his recognition, Lipo shared his story of how and why he joined APEN: "I like to fight for justice because APEN understands the principle that all humans, poor or rich around the world, especially the poor, need justice."[6]

Another inspiring example is Zhi Long Zhang (see figure 5-3), one of our Chinatown leaders, who was recognized with Oakland Rising's 2014 Townie Volunteer of the Year Award.[7] He was selected for this honor because of his leadership and commitment to building power and improving the lives of Asian American immigrants. As he once told a reporter, "I joined APEN because I believe in their mission to improve our communities and have us live in a clean and healthy environment where people can play and work and live and is inclusive of all ages."[8]

Challenges

As with any effort to affect change, building a powerful movement, organizing people, and advancing leadership development have their challenges. Many of us have very little understanding as to how energy gets to the outlet where we plug in our cell phones or to the gas pump where we fill up our cars. For people faced with health problems, rising rents, and unemployment, energy democracy may not be a priority. Even for frontline community members who know very well the pollution and the extraction happening in their own backyards and the resulting impacts on their bodies still may not be familiar with the laws pertaining to that pollution. The arcane world of energy policy can seem impenetrable for the average person and deeply alienating to immigrant communities. That is why deep and ongoing political education is needed to demystify energy policy and systems, bring more awareness and understanding of our current energy economy, and illuminate possible alternatives.

Additionally, the process to bring about deep and systemic change is a process that happens over the long term. Sometimes the changes are not felt until the accumulation of those impacts over time suddenly results in a visible transformation. There is a need for long-term investment and commitment. The process of transformation is iterative and influenced by transformation happening in other places. For example, transformation happening in individuals can spur transformation of the collective community, which can then catalyze changes in the system. Similarly, successful campaigns in transforming the system can also have a profound impact on transforming individuals in that process. Unfortunately, this type of deep investment is not always supported, adequately resourced, or recognized by other groups.

Another challenge is overcoming the dominant narrative and systems of oppression that have told people of color or women that leaders don't look like them. At APEN, our membership reflects intersections of race, class, and gender. Among our local membership are four languages and dialects, different generations, different cultures, different countries of origin, different educational backgrounds, and so forth. The work to support people to be leaders may look very different from person to person.

Many of our members feel intimidated by having to speak in front of a group simply because they don't know English well, especially when having experienced intimidation and harassment from a hostile public. A group of women members who never had a chance to go to school may face challenges in feeling they can voice their opinions in a society that values those with advanced degrees as "experts." Some younger members may feel hesitant to disagree with opinions of older members out of respect for their community leaders. Others may have come from homeland experiences in which organizing and advocacy were not allowed. Some feel unsure about how to make campaign or organizational direction decisions when they haven't had the opportunity to do so before. Many express frustration that the wealthy and powerful have more influence on public processes and elected officials. All of these dynamics make the organizing and leadership development work all the more challenging and absolutely critical.

Conclusion

Given the complexity of our current systems, we know the journey to a regenerative economy will not be easy, and we know that building a powerful movement centering the leadership of frontline communities is required to achieve it. The experiences of APEN illustrate that building power occurs through the organizing and development of people. Whether it is opportunities to distribute ownership or engage in real governance, the new energy economy we're building will require new systems, new practices, and new ways to orient development of our full potential.

What our communities need so intensely right now are real examples of a vision for a new culture and a new society right in our own neighborhoods. We need to be able to touch and feel and experience them and also experience building them as participants in our democracy.

Building new models of energy democracy is part of staying hopeful about our ability to control our own destiny and fight for alternatives to the current systems, which are leaving people vulnerable, undervalued, and unable to meet their families' basic needs.

Ultimately, through organizing for energy democracy, frontline communities are fighting for and winning our right to live in healthy and safe neighborhoods, to have jobs that support our families, to build local economies that serve the real needs of working people, to make our own decisions and create our own solutions, to speak our languages and be proud of our cultures, and to imagine and build the thriving communities we all want for our future.

1. California Environmental Justice Alliance, "CEJA Leaders Show Statewide Support for SB 350," video, https://youtube/2joFsiS6sFQ, 2015.
2. Statement of the Asian Pacific Environmental Network, True Cost of Chevron Report, 2012, http://truecostofchevron.com/assets/docs/2012-richmond-apen-statement.pdf.
3. Ibid.
4. Vivian Huang, "Building Transit-Oriented Community in Oakland Chinatown," *Race, Poverty, & the Environment* 18, no. 1 (2011).
5. The White House, Champions of Change awards, https://obamawhitehouse.archives.gov /champions/community-resilience-leaders.
6. This video is not available online.
7. Oakland Rising, June 23, 2014, http://www.oaklandrising.org/blog/you-got-down-opp.
8. KTVU has eliminated the interview from its website, but there's a blurb about it on the APEN website, http://apen4ej.org/apen-leaders-run/.

CHAPTER 6

Organizing for Energy Democracy in Rural Electric Cooperatives

DERRICK JOHNSON AND ASHURA LEWIS

The small crowd sat mostly still, sharing quiet whispers of conversation, creating a general soft hum about the room. It had been a long but exciting day, and the group remained long after the session had ended, exchanging their thoughts and swapping their own tales of surprise. Interspersed throughout all the subtle mutterings was the repeated exclamation, "I didn't know I owned an electric company!"

This was the scene at the One Voice Electric Cooperative Leadership Institute (ECLI) opening session in July 2016. Over three dozen individuals from different cooperatives had come together on the Tougaloo College campus in Jackson, Mississippi, to learn more about the cooperative that powered their lives. ECLI was developed as a recurring, six-month intensive training program to facilitate the realization of participatory power by co-op member-owners. The single greatest revelation of these One Voice trainings continues to be that participant members are not just customers—they are owners. For these member-owners, ECLI is a first step on the path to creating a true, participatory energy democracy.

ECLI is the brainchild of One Voice, a nonprofit, civic engagement organization based in Mississippi that is focused on building and supporting the active participation of southern black and minority communities across economic, environmental, and climate justice movements. ECLI manifests this vision through its strategy to organize vulnerable and historically racialized

93

communities toward energy democracy. Energy democratization as a mission and electric co-op organizing as a strategy deepen a unique intersection between civic participation, climate justice, and community empowerment. What the general public has not known is that energy companies (cooperative or otherwise) are a major catalyst for economic development and are in a position to "certify" megasites for large business developments and other projects seeking to locate in their service areas. As member-owners of electric cooperatives, communities have an opportunity to address historical, systemic economic insecurity and persistent poverty by reframing how economic growth and innovation take place, at both the level of individual households and the level of community agency.

Through accountable governance structures, electric cooperatives can reduce or eliminate the burning of fossil fuels by implementing more renewable options such as solar and wind power generation, and they can also institute needed energy cost savings through carbon-neutral home retrofits. Using community engagement models like ECLI, One Voice deepens its commitment to black and minority communities and promotes a narrative that makes black and brown liberation a multidimensional story about energy, institutions, race, and access.

The Institute for Local Energy Reliance defines energy democracy as "an energy system that is democratic, where decisions are made by the users of energy."[1] The idea is as elegant as it is simple. In a democratic society, people should be in control of all the major aspects of their lives, and nothing is more major or more fundamental to modern living than electricity. Although currently most energy sources are controlled, and in many cases monopolized, by large corporate entities, it is the ideal and purpose of energy democracy initiatives to decentralize that control and to place authority for decision making and innovation into the hands of the people who use the energy. In an energy democracy, people are free to direct resources toward creating cleaner/greener and more-efficient production and distribution technologies. In addition to making energy sources more adaptive to the needs of the communities that use them, energy democracy also facilitates greater financial stability and serves as a powerful vehicle for job creation and economic stimulation.

Unfortunately, few people live in communities where energy democracy is practiced. Indeed, many communities experience quite the opposite. In a 2016 paper on energy democracy, J. J. McMurtry, associate professor of business at York University, Canada, summarized the existing dynamic of energy resource control as follows:

> While the transition away from fossil-based resources is an important component of the fight against climate change, what is often overlooked is the centralized ownership and control of electricity generation by corporate and state actors. This ownership scheme overwhelmingly favors electricity generation for the sake of profit and growth instead of human and ecological realities. Meanwhile, those who are most directly impacted by the destructive elements of the electricity sector, namely community members and workers worldwide, are excluded from ownership and circles of decision-making. This lack of democracy in the electricity sector is mutually reinforcing with a lack of democracy in the economic and political realms produced and reproduced daily by capitalistic social relations.[2]

It is the lack of ownership and control of energy resources that keeps impoverished communities in states of perpetual economic crisis and instability. For example, in Belzoni, Mississippi, when the average annual family income is compared with the average annual family cost of electricity from the cooperative, the cost of electricity can reach as high as 42% of family income. This is directly caused by the miseducation of member-owners as to their rights and the perceived lack of ownership and control over energy resources created as a result.

However, the situation in Belzoni is not unique to Mississippi or the South. It is a problem suffered across the nation in communities that have high poverty, high populations of people of color, or both. For example, in 2015, New York State enacted the Shared Renewables initiative with the promise that it would ensure that "all New Yorkers, regardless of their zip code or income, have the opportunity to access clean and affordable power."[3] However, in reality, the policy does little to assist impoverished and low-wealth communities achieve the lower bills, increased job opportunities, and generally healthier environment that control over energy policy could bring.

Many states share the same paradigm of power in which it is always the poorest and most vulnerable communities that suffer. The key to bridging the economic gap is the *decentralization* of energy—the transition of authority to the community itself. Only under the stewardship of the communities being served can the goals of energy democracy be achieved, namely to resist the dominant agenda of the large energy corporations and their allies; to restructure energy generation and delivery in order to scale up renewable and low-carbon energy options for low-income households; to ensure energy-based job creation and local wealth creation; and to continuously build community and democratic control over local energy resources.

The History of Electric Cooperatives

Only a few years before the start of World War I, as many as 90% of rural homes lacked electricity.[4] The course of the sun still dictated activities, as, after sunset, farmers were limited to the weak illumination of lantern light. Homes were still heated by fireplaces, and laundry was done by hand. The absence of electricity not only limited physical work activities but also limited economic opportunities, as residents of rural areas were strictly confined to agriculture. Furthermore, businesses and factories required large amounts of electrical power and therefore were not interested in locating facilities in rural areas that lacked the electrical infrastructure they required.

It was only after the federal government took the initiative to address the lack of infrastructure that the electrification of small towns and villages became possible.

In 1933, the federal government authorized the Tennessee Valley Authority to begin building the grid necessary to bring electricity to rural homes in Tennessee. Two years later, in 1935, on the heels of the Great Depression and as part of the New Deal, President Franklin D. Roosevelt signed the executive order that created the Rural Electrification Administration, or REA.[5] Passed a year later, the act that created the REA hit a snag as it soon became evident that few companies were interested in investing substantial capital into electrifying sparsely populated areas despite federal incentives. However, prospects for the act were buoyed when a group of investors did show interest in such an investment. The interested parties turned out to be the farmers themselves.

With so many farm-based applications, it became clear that the best path to rural electrification was through cooperatives.[6] As a result of this revelation, the REA laid the foundation in creating the Electric Cooperative Corporation Act in 1937, which served as a model for states to enact their own laws establishing and governing the creation of nonprofit, member-owned electric cooperatives.

The results were immediate and astounding. The organization America's Electric Cooperatives states that "within four years following the close of World War II, the number of rural electric systems in operation doubled, the number of consumers connected more than tripled and the miles of energized line grew more than five-fold. By 1953, more than 90 percent of U.S. farms had electricity."[7]

In 2015, the number of rural homes receiving electricity was nearly 100%, with the vast majority being powered by *community-owned electric cooperatives.*[8]

Created to be more than simply energy providers, electric cooperatives were intended to operate around seven core principles:(1) voluntary and open membership; (2) democratic member control; (3) members' economic participation; (4) autonomy and independence; (5) education, training, and information; (6) cooperation among cooperatives; and (7) concern for community.[9] As the principles indicate, at their very foundation electric cooperatives were designed to be democratic institutions. As time progressed, many electric cooperatives began to behave more like their impersonal, corporate, for-profit utility counterparts and less like the community empowerment vehicles they were created to be. For example, one investigation "routinely found excessive rates, inefficient operations, punitive collection policies, padded executive salaries, and discrimination in co-op investments and hiring."[10]

As of 2015, there were 930 electric cooperatives throughout the United States. Of those, 864 distributed electricity, accounting for 10% of the nation's annual total kilowatt-hour usage. They serviced 42 million customers in forty-seven states, accounting for roughly 12% of the nation's total electricity market.[11] These distribution cooperatives also owned and maintained 42% of the nation's distribution lines. The remaining 66 cooperatives were generation and transmission (G&T) cooperatives.[12] In addition, these 930 cooperatives serviced over 90% of the nation's persistent poverty counties.

The role of rural electric cooperatives in Mississippi is equally significant: In 2015, Mississippi's 25 distribution co-ops and one G&T co-op provided *nearly half the electricity in the state* and serviced 98% of the persistent poverty counties in the state.[13]

Energy Democracy as a Pathway Out of Poverty

As mentioned earlier, electric cooperatives were established in the 1930s at a time when few African Americans were fortunate enough to own land or allowed to exercise that ownership without threat of violence. Existing as second-class citizens, African Americans were not generally in a position to enjoy the benefits of cooperative participation. To make matters worse, those who *were* members of an electric cooperative were deliberately kept in the dark about their member-owner rights and were systematically excluded from any participation in voting, once again under the threat of violence.

This situation has changed form and, in many cases, become more subtle, but the effects remain. Most African American member-owners of electric cooperatives are unaware of their rights, and lack of information discourages participation in democratic voting. Also, just like the cooperatives of almost a century ago, the decision-making power of the board of directors sits almost exclusively in white hands.

There is no contesting that the mostly African American communities of the state of Mississippi still suffer from a legacy of racism and discrimination that has led to their current poverty. Economic stagnation and racially disparate concentrations of wealth and power have resulted in an overwhelming sense of frustration and hopelessness among many community members. Though the establishment of electric cooperatives in these communities should have relieved some of the economic stresses, in reality many of the cooperatives have in fact done just the opposite.

Poor governance by the boards of directors of the cooperatives has contributed to economic slowdowns, the stifling of innovation, and the erosion of occupational opportunities. Instead of encouraging democratic elections, cooperative boards began to homogenize in ways that did not reflect the membership. Instead of sharing power, cooperative boards began to consolidate control and vest all decision making in a select few. Instead of providing economic security for the community, cooperative boards began to take advantage of the state's minimal authority to regulate them and started charging exorbitant rates to the membership while becoming less and less transparent about cooperative finances.

In 2014, Mississippi's twenty-six electric cooperatives reported a total worth of over $5.2 billion and a total profit of $161 million. As nonprofit entities, electric cooperatives must return all profits to the member-owners; however, only 30% of Mississippi's electric cooperatives returned any profits to their member-owners. To make matters worse, Mississippi areas serviced by municipal or private energy institutions typically have higher average wealth yet pay significantly less for electricity than their poorer cooperative counterparts. In fact, in some extreme instances, cooperative members spend up to 42% of their total income on electric service, this in a state with a 28% lower household income than the national average.[14] The disparity is worse when comparing census data against the makeup of the boards of directors of cooperatives in Mississippi.

While the state is 37% African American and 52% female, board representation for Mississippi's electric cooperatives comes in at an astonishingly low 6.6% African American and 4% female, a direct result of undemocratic elections. It is precisely this corrosive discrimination that the ideal of energy democracy seeks to address.

Placing power and control over the direction of electric cooperatives in the hands of those serviced by them establishes real ownership. True community control takes root. Former U.S. secretary of labor Robert

Reich recently made this point at the 2016 Quebec International Summit of Cooperatives, stating, "A cooperative takes the ideal of democracy and puts it into our workplaces and our organizations. And cooperatives ideally could be a countervailing power to the large forces of large corporations and the biggest banks and the biggest institutions."[15]

Truly democratic elections encourage member-owner buy-ins that can result in a board of directors that is more reflective of the demographics within each cooperative. Transparent finances and appropriate remuneration to member-owners can provide and promote financial stability. Community ownership and inclusion facilitates innovation, which, in turn, paves the way for new job opportunities and economic expansion. To put it simply: democratizing the electric cooperatives is a pathway out of poverty in Mississippi.

Lessons Learned: A Brief History of Mississippi Organizing

In Mississippi, the African American community is well versed in the practice of resistance. From informal black Civil War "contraband" militia and the many divisions of Union Colored Troops to World War II black soldiers, pilots, and naval officers, African Americans in general and specifically from Mississippi have never sat quietly by, indulging in indifference while suffering under an oppressive yoke of diminished rights, human or civil.

While an entire book could be dedicated to the centuries of struggle endured by Mississippians of color, this discussion will be limited, starting with the return of World War II veterans, focusing specifically on the type of organizing strategies that have developed since then and why they were successful.

At the close of World War II, African American veterans found themselves returning to a state that neither recognized nor allowed the expression of the very rights they risked life and limb on foreign shores to protect. However, instead of fleeing the state in search of greener, and more equitable, pastures, these veterans remained and formed a loose but effective network that became the foundation of the National Association for the Advancement of Colored People—the NAACP—in Mississippi.

The network was so effective that it was instrumental in demanding the investigation of the 1955 murder of Emmett Till, a fourteen-year-old black boy falsely accused of flirting with a white woman. It was, in fact, the worldwide coverage of this atrocity that brought attention to the violence and fear that many African Americans faced on a daily basis and placed an uncomfortable spotlight on the state of Mississippi and the perpetrators of violence. As tragic and appalling as Till's murder was, it also served to light a spark in young people, many of whom were Emmett's age.

The rise of the Student Nonviolent Coordinating Committee (SNCC), working in tandem with the NAACP chapters (led primarily by World War II veterans), formalized the statewide network and connections still used today. Though never short of powerful leaders (later turned icons of the movement: Aaron Henry, Amzie Moore, Medgar Evers, Rev. T. R. M. How-ard, Hollis Watkins, and many, many others), organizing efforts in Missis-sippi chose to follow the philosophy so eloquently phrased by Ella Baker: "Strong people don't need strong leaders." It was this philosophy that took Mississippi from the path of an "egocentric" organizing model adopted by other southern states to a "community-centric" organizing model.

One Voice defines community-centric organizing as the ability of the people who are most directly affected by an issue to leverage their collec-tive capacities to implement a solution-based strategy. It is not dependent on any individual's personality or on one powerful, charismatic leader but relies instead on the collective strength and wisdom of the commu-nity to confront the injustices to be addressed. When empowered with the information, tools, and resources to enable change, communities will self-direct their course in altering their condition.

In contrast, an "ego-centric" organizing model is one in which all organizing efforts flow around and through one or a few powerful, charismatic leaders. While One Voice does not criticize the ego-centric approach, the community-centric model has had a greater history of suc-cess for Mississippians.

In the 1960s and '70s, community-centric organizing enabled whole communities to strategize, navigate, and focus their energies, result-ing in much greater success than efforts in other southern states. The community-centric approach led to the successful holding of Freedom Summer in 1964, which in turn provided the documentation necessary for the passage of the 1965 Voting Rights Act. The enormous voter turnout in Mississippi as a result of the act paved the way for the strategic redistricting that followed, drawing political boundary lines in such a way that allowed more communities to elect candidates of their choice who represented their interests. As of 2017 and as a direct result of this approach, Mississippi has more African American elected officials than any other state in the nation. This was all made possible because of a stalwart belief in a simple truth: affected communities are in the best position to determine where their interests lie and what their needs are.

With the lessons of the past firmly in mind, this same philosophy is utilized in all of One Voice's work. Under the tutelage of civil rights movement icon Hollis Watkins (figure 6-1), One Voice's strategy in many areas (including its work with electric cooperatives) is guided by four tried-and-true steps to successful community-centric organizing: investigation, education, negotiation, and demonstration.

Though the steps are to be seen through in order, it is important to realize that each step is not mutually exclusive of the others. At several points during the organizing process, the organizer or organization may refer to a previous step or conduct two or more steps simultaneously. What's more, some steps may never truly end until the entire organizing effort has concluded. This interconnectivity between the steps themselves is the key to the process's success as it allows for adaptability, assessment, and reflection in real time. For example, the process of investigating never ends. As new information becomes available, the organizer must update existing information and educate the community on new developments.

Now for a closer look at each step:

Investigation is the first step in the process and, as the name implies, simply means that prior to any actual organization, research must be done. Before a problem can be addressed or people can be organized, the reality of the situation has to be firmly established. It is important that all assumptions are recognized and banished quickly in the beginning. Ignoring or maintaining them will only skew and ultimately hinder any

Figure 6-1. Civil rights icon Hollis Watkins often leads the One Voice staff in organizing train-ings and has served as a constant fount of knowledge since the organization's inception. Here, he explains the four steps of community-centric organizing. *Image source: One Voice*

real work from being accomplished. Also, the investigation is not purely an academic exercise. Though printed materials and statistical data can be useful in painting a picture, the full image will not be complete until it includes the experiences and perceptions of the people involved. There-fore, it is essential that the investigation phase include speaking with those who are the subjects of the organizing attempt. Once the prelim-inary investigation has been completed (it is preliminary because it is highly likely that further investigation will be warranted throughout the organizing process), it is time to transition into the second step, education.

The **education** step addresses the need to increase the base knowledge of all those involved around a particular subject. Here is another place where acting on assumptions will ruin an organizing attempt. People often equate levels of formal education with knowledge or intelligence.

However, experience has repeatedly shown that this is not the case. The man with an eighth-grade education who runs the local barber shop may turn out to be the most knowledgeable person in the community. He may know how best to reach the most members of the community at the same time or which church leader has the most sway over the members. It has been said that an expert is the person in the room who knows the most about the subject at hand. This should be kept in mind in both the investigation and education phases. Complementing the investigation phase, in which the organizer becomes educated about the issues and obstacles, the education phase educates the community about the many layers of context surrounding the problems and their solutions across policy-based, legal, political, technological, and financial fronts.

The word **negotiation** may bring to mind two opponents glaring at one another across a conference table, but as part of the organizing process, it refers to the subtle yet intricate task of ensuring all parties are on the same page and working toward the same goal. Internal negotiations are conducted among those seeking to organize the community. External negotiations are done between the organizers and the community they are attempting to organize. Though everyone at the "table" should be allies, that does not mean all will be in agreement with any particular assessments or plans of action. As a result, there must be internal as well as external negotiations among those involved—as with investigation, this step is likely to continue throughout the organizing process.

The final step is **demonstration**. In the demonstration phase of the process, the plan of action created, based on the results of the multiple investigations and negotiations, is finally carried out by an educated and knowledgeable community.

Using these steps and the approach that follows, One Voice's work with electric cooperatives and community leadership is not only capable of producing tangible and lasting results, it is also capable of replication. Though each state may have slight differences and community idiosyncrasies, there is a shared history throughout the southern region that makes the One Voice approach easily applicable to many different individual cooperative circumstances while still retaining a high degree of adaptability. In this way, One Voice's work on energy democracy through rural electric cooperatives can serve as a relevant case study for other environmental justice or energy activists and organizations.

One Voice's Organizing Campaign

The goal of One Voice's Energy Democracy Initiative and the Electric Cooperative Leadership Institute is simple: to bring energy democracy to Mississippi's electric cooperatives and the members they serve. To accomplish this end, One Voice's energy initiative has three major phases:

- **Phase 1: Listening.** Starting in spring 2015 (and continuing throughout the life of the project), the first phase was the foundation of the energy campaign in its entirety. Phase 1 consisted of listening tours by One Voice staff to all the electric cooperatives in the state. Community members shared with One Voice their concerns regarding the lack of transparency and the questionable practices found within their cooperatives. During this phase, general information was shared with the members, explaining their dual roles as member-owners, but, admittedly, without significant depth. That came with phase 2.
- **Phase 2: Creating ECLI.** Begun in the summer of 2016, the second phase involved the creation of the Electric Cooperative Leadership Institute (ECLI) and individual campaign development by the members. A cohort composed of a cross-section of member-owners across all the target cooperatives was introduced to the major concepts and principles around the rights and roles of their cooperatives and the basics of community-centric organizing. The cohort then created their electric cooperative–specific education campaign that would be used to bring the same information to the entire cooperative membership.
- **Phase 3: Election Campaigns.** One Voice is now, in 2017, in the third and final phase, which was built on the first two phases' outcome of increased participation among members across their cooperatives. This phase includes candidate selection and election campaigns intended to result in more diverse board representation within the

different cooperatives. In this phase, One Voice is taking a less direct role, leaving the community and the cooperative membership to make the organizing decisions while providing them with the tools, support, resources, and technical assistance necessary to make their goals a reality.

THE ELECTRIC COOPERATIVE LEADERSHIP INSTITUTE (ECLI)

For One Voice, the most important step toward democratizing Mississippi's electric cooperatives is member education. One Voice's Electric Cooperative Leadership Institute was developed as a multisession, six-month-long, in-depth member education program to equip members with financial knowledge, best practices, and community organizing tools to take back to their cooperatives. Participants in the education program strategize with one another on how to best educate their entire membership bases, and they continuously research innovative ways in which other electric cooperatives are providing economic opportunity to their members. Through this pathway to change, communities can transform their electric cooperatives into true member-driven institutions that provide a crucial vehicle for each community's economic development, empowerment, and financial security.

THE ECLI CURRICULUM

The curriculum One Voice has developed to support the ECLI training sessions covers several topics, but a few will be described here: the concept of *member ownership* of the cooperative, the importance of knowing and operating within the *bylaws* of the cooperative, and understanding *cooperative finances*.

Within the context of *member ownership*, members need to understand that their cooperative exists as a utility that can provide services at cost or at a lower cost than other comparable electricity providers. Consumers who are serviced by for-profit providers pay their bills and receive electricity, but that concludes the exchange. In contrast, One Voice wants ECLI participants to understand that as member-owners, their relationship to their service provider does not mimic the for-profit interaction. Instead, as a member-owner, the mere fact that they receive power from their cooperative awards them membership in the cooperative, and their membership in turn affords them a stake and degree of ownership in the cooperative itself. This means that, as recipients of services, they are not passive consumers but can affect the daily operations and programs of the cooperative that have a direct impact on them and the greater community.

A significant portion of ECLI sessions focuses on presenting and explaining each cooperative's *bylaws*. The cooperative's bylaws are the governing documents regarding how the cooperative is to function. Whether to implement change, increase board diversity, or just be more engaged members, member-owners must maintain a fundamental understanding not only of their bylaws, but the membership's relation to their bylaws. Unengaged members, for example, do not attend meetings and, as a result, cannot vote either for or against proposed changes to the bylaws, depending on whether they are or are not in the best interests of the members. Ignorance of bylaws can also greatly influence the outcome of elections. If a member is not familiar with election procedures, such as the number of members that constitute a quorum or acceptable proxy votes, then any strategy or plan to increase board diversity, for example, through supporting specific candidates is at high risk of failure.

An example of how democratic change in governance works well can be found with Baton Rouge's Dixie Electric Membership Cooperative. In 1983, an interracial group of member-owners staged a "democratic revolt" that resulted in the election of a completely new board of directors. Following the revolt, the member-owners have seen their electric bills decrease and their economic opportunities increase.[16] By contrast, when member-owners of Mississippi's Delta Electric Power Association attempted a similar revolt, the all-white board of directors simply left the room, waited a few months, and reelected themselves in secret.[17] To this day, neither the board nor its actions have been challenged, as member-owners have not realized their power, and the board is still protected by the language of the cooperative's bylaws regarding election procedures and quorums.

Another major facet of the ECLI curriculum includes ensuring that the member-owners have a functional understanding of *cooperative finances.* It needs to be clear in the minds of member-owners whether their cooperative is operating in a manner that is in compliance with its bylaws as well as relevant applicable outside laws and regulatory agencies. A memorandum sent from "Interested Parties" to Brandon Presley, public service commissioner for Mississippi's Northern District, stated:

> There has been additional anecdotal evidence from customers and industry representatives which indicates concern over the lack of capital credit returns to memberships. There appears to be evidence of nationwide non-compliance as well as likely non-compliance with Mississippi law by Mississippi EPAs.[18]

In addition, member-owners should not only be aware that funds in excess of expenses either should be used to increase the efficiency of services and infrastructure or returned to the membership in the form of patronage capital; they should also be aware that if the funds are not used in this manner, they are entitled to know exactly how the funds have been or will be used. A joint report of the National Rural Electric Cooperative Association and the National Rural Electric Cooperative Finance Corporation explains patronage capital as follows:

> [A]ny excess of operative revenues collected over operating expenses from the provision of electricity must be allocated to patrons as capital credits, based on their participation, and ultimately returned to patrons. . . . Capital credits should be accounted for in a way that reflects the rights and interests of members in the net savings of the cooperative. These rights and interests must be protected and not forfeited.[19]

These topics, as well as many others, are addressed throughout the course of the ECLI session, giving member-owners the knowledge base to create multidimensional strategies for becoming more actively engaged in the governance of their respective cooperatives.

ECLI SESSIONS

At the ECLI inaugural session, One Voice's facilitators noted one thing immediately: when the participants arrived, they walked in with a purpose. Though there was some trepidation evident in the small smiles and quiet movements as the members drifted toward their seats, it was clear that most were more eager than anxious as they waited for the session to begin, notepads flipped to fresh sheets and pens at the ready to engage with the topic at hand. Group discussions laid the foundation for a shared bond, vision, and values associated with an authentic cooperative enterprise (figure 6-2).

As the session progressed, not only did participants feel more at ease, their desire to know more about their cooperative grew in counterpoint to any lingering anxiety. The discussion of cooperative bylaws was particularly energetic as members relayed their personal experiences in relation to their board's exclusionary tactics. One member was told her membership would be suspended if she made a late payment on her electric bill. Another stated her cooperative told her a credit check was required before permitting membership. Similar stories of intimidation, misinformation,

Figure 6-2. Participants exchange thoughts at an Electric Cooperative Leadership session.
Image source: One Voice

and, in some instances, blatant untruths were shared throughout the morning. However, the feelings of frustration and embarrassment layered within these experiences were later transmuted into a fierce desire to share what they had learned once they had examined their actual bylaws—namely, that their cooperatives were treating them wrongly.

And this enthusiasm never wavered. Whether the topic was ownership, legal standing and authorities, or election procedures, the members were attentive and engaged. Even the hours spent sorting through the complicated maze of tax forms, revenues, and regulations did nothing to dissuade the stunning alacrity displayed by the members, nor did it diminish the voracious appetite they displayed when encountering new knowledge about their cooperatives.

Perhaps the most dynamic session was the day that members spent with Hollis Watkins, learning for themselves the fundamentals of organizing and the steps to make it successful. These lessons were delivered in an innovative and wholly interactive manner, as members were able to actually try their hand at collecting the proxy votes necessary for a fictional cooperative election day via a card-based role-playing game.

However, One Voice's facilitators found that, for the organization, one of the most rewarding experiences throughout the ECLI sessions was that moment when the members moved from day one's "I didn't know I owned a cooperative" to "here are the policies I want my co-op to adopt." At the start of the institute, the class of member-owners spoke very openly about their lack of knowledge regarding their cooperative rights, roles, and responsibilities. Six months later, however, the class spoke just as openly, and with a newfound confidence, about the seven cooperative principles, how they should be applied, and what their individual cooperatives could and should do to operate in a manner more aligned with those principles.

The Road Ahead

With increased awareness within the membership and a full understanding of how their cooperative should operate, member-owners will begin to tackle the issues found with their respective boards of directors via candidate recruitment and election campaigns. Successful campaigns will result in an increased presence of both African American and female directors that, in turn, will begin to shift representation in ways that are more reflective of the membership as a whole. ECLI participants have already crafted their strategic plans and, with the guidance and technical support of One Voice, will begin implementing those plans in coordination with their cooperatives' annual meetings.

In addition to cultivating a board of directors that is sensitive to the needs of its cooperative and its membership, member-owners will begin to focus on investigating and implementing best-practice models while exploring renewable energy resources. These options will span from on-bill financing and net metering to comprehensive weatherization and alternative energy sources, along with everything in between. With the freedom to direct the decision making within the leadership of their cooperative, member-owners will enjoy not only the benefits of a responsive board but also a deep sense of ownership and economic stability. These benefits have the potential to grow into increased climate resilience within the community, improved general health status of the community, decreased economic stress, and less dependence on fossil fuels and other toxic energy sources.

One Voice's work continues as it prepares for its next ECLI session, slated to begin in May of 2017. Each cohort will include new representatives from the participating cooperatives as well as representatives from two additional cooperatives, thus expanding the knowledge to more member-owners and strengthening energy democracy for the targeted cooperatives. Also, as the prior participants begin to implement their cooperative-specific strategic plans, One Voice will expand its role from the relatively narrow scope of educator to one of technical assistance and support for those who have already completed the leadership training program. Even more is on the horizon, as the energy campaign will now cross state lines, stretching into Alabama, Arkansas, Louisiana, Tennessee, and Georgia as partner organizations seek to emulate One Voice's approach to expand energy democracy in the South.

1. Institute for Local Self-Reliance, "A New Logo, and a Definition of Energy Democracy," January 12, 2017, https://ilsr.org/a-new-logo-and-a-definition-of-energy-democracy.
2. J. J. McMurtry and Derya Mumtaz Tarhan, "Energy Democracy: A Liberatory Conceptualization," accessed April 20, 2017, http://www.ciriec.ulg.ac.be/wp-content/uploads/2016/10/REIMS-TARHAN-CA.pdf.
3. New York State Energy Democracy Alliance, "EDA Response to the Release of the Report on Low-Income Participation in Shared Solar" (2015), accessed April 20, 2017, http://energy democracyny.org/wp-content/uploads/2015/04/EDA-Response-to-the-Release-of-the -Report-on-Low_final.pdf.
4. America's Electric Cooperatives, "History," accessed January 12, 2017, http://www.electric .coop/our-organization/history.
5. Ibid. *Note:* The REA later became the Rural Utilities Service (RUS) and falls under the auspices of the United States Department of Agriculture.
6. Ibid.
7. Ibid.
8. Ibid.
9. America's Electric Cooperatives, "Understanding the Seven Cooperative Principles," accessed December 01, 2016, www.electric.coop/seven-cooperative-principles%E2%80%8B.
10. Henry Leiferman and Pat Wehner, "A Question of Power: Race and Democracy in Rural Electric Co-ops," *Journal of the Southern Regional Council* 18, no. 34 (1996).
11. University of Wisconsin Center for Cooperatives, "Rural Electric Overview," Research on the Economic Impact of Cooperatives, accessed April 20, 2017, www.electric.coop/category /cooperative-advantage/coop-101.
12. Ibid.
13. America's Electric Cooperatives, http://www.electric.coop/category/cooperative-advantage /coop-101.
14. Carmen DeNava-Walt and Bernadette Proctor, "Income and Poverty in the United States: 2014" (Census Library Publications, 2015), accessed January 09, 2017, www.census.gov /content/dam/Census/library/publications/2015/demo/p60-252.pdf.
15. Steven Johnson, "Robert Reich on Co-ops," *America's Electric Cooperatives* (October 23, 2016), accessed April 20, 2017, www.electric.coop/robert-reich-on-cooperatives-international -summit/.
16. Leiferman and Wehner, "A Question of Power" (see n. 10).
17. Ibid.
18. "Internal Memorandum: Non-compliance," to Brandon Presley from Interested Parties, Re: Initial Inquiry into MPA Capital Credits Policies, September 2014.
19. National Rural Electric Cooperative Association and the National Rural Cooperative Finance Corporation, *Capital Credit Task Force Report*, ch. 2, p. 24 (2005).

CHAPTER 7

Conflicting Agendas: Energy Democracy and the Labor Movement

SEAN SWEENEY

Energy is today at the center of huge social and political conflicts in many parts of the world, and the number of struggles appears to be proliferating. Unions are engaging in these struggles, but often on different sides. Some have sided with their employers in the coal, oil, and gas industries; meanwhile, others—driven by concerns about air and water pollution, land despoliation and seizures, displacement of communities, climate change, and so forth—are part of the opposition to the expansion of fossil fuels and "extractivism" in general. The outcome of present-day struggles will have a clear bearing on the effort to democratize energy in the years ahead.

This chapter will summarize how three broad but distinct agendas today compete for some kind of dominance on the global stage. It will then describe how these agendas are expressing themselves in labor struggles in the United States. It will describe how, in both the United States and in the global labor movement, energy democracy became a distinct component of a "system change" narrative that is in the ascendancy but still far from being dominant. Last, it will identify some of the challenges facing the energy democracy current in the international labor movement and how these might be confronted and resolved in the period ahead.

Agenda Wars

At the level of ideology and narrative, three distinct "agendas" presently compete with one another at the global level.

First, there is what can be described as the fossil fuel or "business as usual" (BAU) agenda. In 2012, companies and utilities represented nineteen of the world's fifty leading corporations. They accounted for 48% of the revenues and nearly 46% of the profits of the top-fifty group.[1] Roughly 30 million workers are thought to be active in this sector. Fossil fuel interests broadly agree that fossil fuels should and will be the leading source of energy in the decades to come.

The second agenda is "green growth," a concept that was elaborated in the landmark publication *The Economics of Climate Change: The Stern Review*, released in 2006.[2] Green growth is anchored in ideas of "ecological modernization," which envisages a fully decarbonized future phase of capitalism. A term rarely heard before 2008, green growth has come to occupy a prominent position in the policy discourse at the international level.

The central tenet of green growth is that environmental problems are caused by "externalities" like greenhouse gases (GHGs). Once these externalities have been priced appropriately (the "polluter pays" principle), investment and business patterns will shift accordingly—for example, driving investment toward a transition to renewable energy—leading to a point at which economic growth can be *decoupled* from environmental damage. The higher the price on externalities like carbon, the sooner the moment of decoupling will arrive. Therefore, capitalist growth (now green) can continue more or less as it has done for the past two or more centuries.

The third agenda (more accurately, an "agenda in formation") is presently without a distinct name, but it incorporates the politics of "system change." It challenges the very idea that capitalism might be capable of decoupling growth from emissions and other environmental damage or that there can be, even in theory, some version of "steady state" capitalism that more or less constrains the political economy within the confines of planetary limits. In this view, radical restructuring of political economy is necessary in order to stay within these ecosystem boundaries.

Naomi Klein's book *This Changes Everything: Capitalism vs. the Climate* represents both a synthesis and a powerful popularization of the idea that, in the case of climate change, the problem is not carbon, but "growth without end" capitalism. The "growth imperative" is part of the DNA of capitalism, and attempts to decouple growth from ecological damage will fail and have, indeed, already failed.

Energy democracy has taken root among unions that understand that a radical restructuring of the global political economy is necessary for civilizational survival. If the struggle for energy democracy has "fronts," then organized labor is a particularly important one. As a global institution, unions represent roughly 200 million workers outside of China (where 150 million workers are registered as union members). Unions have a relatively strong presence in the energy sector in many countries. The International Trade Union Confederation (ITUC) alone represents 180 million members and has affiliates (national centers) based in 162 countries. The total number of individual unions linked to these national centers runs into many thousands.

At the level of real politics and social struggles at the global level, unions are clearly significant "players" in the pursuit of all three agendas described. This is clearly seen in the United States, where the pursuit of these agendas has led to increasingly sharp political divisions in the labor movement.

Labor's Stance

The main outlines of these differences in U.S. labor will be sketched out below. The last decade or so has seen the rapid rise, and then the unspectacular fall, of "green growth" unionism in the United States. The resurgence of a BAU agenda based on exporting carbon has clearly won the active support of a wing of U.S. labor. More recently, a gradual cohering of union forces is prepared to challenge the BAU agenda and, potentially, provide a labor platform that is either grounded in the idea of energy democracy or, at the very least, has energy democracy as an important component.

GREEN GROWTH AND THE BLUEGREEN ALLIANCE

The support of U.S. labor for the green growth agenda has largely been framed by the BlueGreen Alliance (BGA), which, at the height of its influence in 2007–2012, was a coalition of thirteen labor unions and several key environmental organizations led by the United Steelworkers and the Sierra Club. The BGA positioned itself in a way that would reinforce the efforts of the liberal wing of the Democratic Party to pursue a green industrial policy agenda directed toward restoring U.S. competitiveness (particularly in manufacturing) through modernization, energy efficiency, upgraded infrastructure, and the deployment of large-scale, centralized renewable power.

The BGA's 2007 launch occurred at a time when the global discourse around climate change had taken center stage following the release of the *Stern Review on the Economics of Climate Change*; the *Fourth Assessment Report (FAR) of the Intergovernmental Panel on Climate Change* (IPCC); and Al Gore's documentary film *An Inconvenient Truth*. The IPCC's report and Gore's documentary communicated the urgency of the climate crisis, while Stern's message also gave a clear sense that the transition to a low-carbon economy was inevitable. The science was, after all, definitive, and a broad consensus was emerging among business, governments, and the rest of "civil society" (unions are included here) that emissions reductions were urgently needed. In addition, the *Stern Review* had shown how the economic transition could bring many benefits (including jobs) and eventually deliver "cleaner and better growth" for decades to come.

By advocating the greening of several key economic sectors, BGA created space for more progressive unions like the Service Employees International Union (SEIU) and the Communications Workers of America (CWA), among others, to engage.[3] These unions mostly went along with the green growth (and green jobs) messages of the large environmental organizations that were active in BGA. The confidence these environmentalists had in the "commonsense" nature of market-led solutions proposed by Stern, the World Bank, and others went more or less unchallenged in progressive labor circles. Union support for this agenda would, they believed, one day be rewarded in the form of millions of "green jobs."

Following the financial crisis of 2008 and the subsequent recession, unions internationally sought to push the green growth agenda into an explicitly job-centered project. They echoed neo-Keynesian calls for the need to restore demand and seize the opportunity to generate "inclusive growth" or "job-centered growth."[4]

Following the arrival of President Obama in the White House in early 2009, the BGA gained considerable prominence and had a major influence over congressional Democrats during the effort to pass national (cap-and-trade) climate legislation that year. Member organizations of BGA were part of a large coalition led by strong "green growth" advocates such as the Environmental Defense Fund (EDF), the Natural Resources Defense Council (NRDC); and the Pew Center on Global Climate Change. The main goal of the coalition, named the United States Climate Action Partnership (USCAP), was to establish a market-based, economy-wide emissions trading system along the lines of the European Union Emissions Trading Scheme, which came into place amid abundant green growth fanfare in 2006.

Political Defeats for Green Growth

In a massive setback for those supporting the green growth agenda in the United States, "coal state" Senate Democrats in mid-2009 sided with Republicans to defeat the climate bill.

Coming on the heels of this defeat, the green growth agenda was dealt another damaging blow in Copenhagen in late 2009, when the United Nations climate talks collapsed and governments failed to negotiate a new climate agreement to succeed the Kyoto Protocol (which was set to expire in 2012). Countries of the global South demanded that the richer countries take the lead by pursuing emissions reductions along the lines advocated by the IPCC.

Anxious to prevent mandatory targets of this nature coming into effect, the United States, China, and other key governments presented a six-page memo called the Copenhagen Accord. The accord sought to establish a voluntary approach to emissions reductions.

The International Trade Union Confederation reacted angrily to the Copenhagen Accord, criticizing among other things, "the lack of ambition in the emission reduction targets pledged by the United States."[5] In contrast to the ITUC's assessment, the report of the American Federation of Labor and Congress of Industrial Organizations (AFL-CIO) from Copenhagen was unreservedly positive. "We found ourselves," stated the AFL-CIO's representative, "closely aligned with the State Department negotiators on issues critical to the U.S."[6]

Notably, BGA also applauded the accord and praised the State Department for its leadership. This gave political expression to a point of tension within the green growth framework, which concerns disagreements between those who emphasize the need for government-led commitments and targets and those who feel the market will ultimately overcome any "lack of political will," as corporations and investors become aware of the returns the green transition will produce.

Most unions around the world took the view that without mandatory emissions reductions and clear targets and timetables to achieve them, green growth would be impeded and the promise of green jobs would remain largely unfulfilled. By contrast, BGA went to Copenhagen determined to support the administration's weak emissions reduction target, thus undermining the green growth agenda at the heart of its own program.

Carbon Markets in Trouble

By this time the main policy mechanism driving the green growth approach, namely carbon pricing, had already fallen into deep trouble. The flagship program, the European Union's Emissions Trading Scheme, had essentially capsized as a result of the recession; low demand for pollution permits; and the generation of too many permits during phase 1 of the scheme, which depressed the carbon price. The idea of a global carbon market to drive green investments and growth, once regarded as politically inevitable by Stern and his cothinkers, now seemed politically untenable.

In 2016, more than twenty years after the Kyoto Protocol had established carbon trading as the principal policy instrument for reducing emissions, still 88% of global GHGs were *not* covered by a price, and where a price exists it is rarely above $10 per ton, making it virtually useless as a tool to reduce emissions. Meanwhile emissions from fossil fuel use rose a staggering 61% in the period from 1990 to 2014.[7]

Both in the United States and globally, the green growth framework is, therefore, in deep trouble at the level of both political strategy and economics. And the failure of the green growth agenda is leading to a serious polarization in the trade union movement.

"SAUDI AMERICA" AND THE BLACK-BLUE ALLIANCE

The political failures of the green growth agenda that occurred in Washington and Copenhagen in the context of a deep recession emboldened the business-as-usual (BAU) agenda of the fossil fuel companies, especially in a number of key countries, including the United States. The BAU agenda rested on the argument that environmental and climate protection measures interfered with much-needed job creation and economic recovery and should be postponed or rejected altogether.

In the United States, Obama's campaign commitment to create 5 million green jobs had not materialized, and U.S. manufacturing took a major hit as a result of the recession. During this period, among the few industries doing well were U.S. oil and gas, buoyed by high global prices. In 2009, the unions of the AFL-CIO's Building and Construction Trades Department (known simply as "the Trades") forged an explicit partnership with the oil and gas industry to develop North American energy sources via the Oil and Gas Industry Labor-Management Committee, which, for convenience, I will describe as the "Black-Blue Alliance"(BBA).[8]

The Trades today represent approximately 2.3 million union members in the United States, roughly one-fifth of the country's 15 million organized workers. Leaders of the Trades do not always see eye to eye, but the Trades' leadership majority had concluded that that the rich coal, oil, and gas resources of the United States would help drag the nation's economy out of the recession, help ensure U.S. economic prosperity in the years ahead, and provide union members in the construction trades with *real* jobs, not imaginary green jobs.

Union support for fossil fuel-driven BAU was therefore strategic, and not simply indicative of a project-by-project foraging for jobs, as is often depicted by the mainstream media. World prices for coal, oil, and gas are presently (2017) very low, but both the Trades and the industry believe that, over the longer term, demand for fossil fuels is assured—unless, of course, climate commitments to reduce emissions get in the way.

Keystone XL and Carbon Exports

An early political test for the BBA was the fight over the Keystone XL Pipeline (KXL). In 2010, four unions signed project labor agreements (PLAs) with the pipeline company, TransCanada Corporation, and the four joined the American Petroleum Institute (API) to lobby then secretary of state Hillary Clinton to issue the pipeline permit.[9] Few at that time doubted that the permit would be issued; the industry wanted it, "labor" wanted it—KXL was a "done deal."[10]

TransCanada, the oil industry, and the unions would use the jobs argument to win a permit to construct the pipeline, and to do the same for a string of future infrastructure projects related to tar oil, coal export terminals, and fracking. Winning an approval for KXL promised union contracts for years to come—a fact that goes a long way toward explaining why KXL became a top priority for construction unions.

Among the more significant features of the fight over KXL was the oil industry's commitment to ensure that U.S. coal, oil, and gas is directed toward overseas markets. With energy consumption in the United States flat or falling, this "rush to the coasts" is one of the drivers of pipelines, coal and gas export terminals, refinery expansions, and the source-to-market movement of fossil fuels in general.

In mid-2015, both the Laborers' International Union of North America (LiUNA) and the International Union of Operating Engineers (IUOE) successfully linked arms with the likes of the Koch-funded[11] Americans for Prosperity, the U.S. Chamber of Commerce, and the API in a call on Congress to lift the export ban on U.S. crude oil that had been introduced in 1975 during the Middle East oil crisis. The two unions stated that "lifting the ban will result in increased domestic crude production and deliver hundreds of thousands of jobs across all sectors of the American economy."[12]

Political Fracturing

The Trades' formal alliance with the API has already changed organized labor's political agenda. In 2015, the North America's Building Trades Unions (NABTU) was established. The new name of the Trades symbolized the growing distance from the AFL-CIO and the intention to forge a closer relationship with industry interests and, it seems, the Republican Party. NABTU announced that it would endorse candidates based on how supportive they were of the kind of industries that employed their members—in effect, a public endorsement of the "export carbon" version of BAU agenda. In an interview, NABTU leader Sean McGarvey stated, "We're not about political parties. We're about construction workers. . . . Both parties have changed, and we have no permanent friends and no permanent enemies."[13]

Following the 2016 presidential election, the Trades quickly warmed up to the agenda of Donald Trump and his key appointees. Responding to the proposed appointment of ExxonMobil CEO Rex Tillerson as secretary of state, NABTU stated, "We believe he will be a tremendous success," praising Tillerson's "resilient and dynamic grasp of both global and domestic policy issues, and a deep and unyielding sense of patriotism for our great nation."[14]

If further evidence were needed to confirm the political direction of the Trades, in late December 2016 the Laborers' Union criticized outgoing president Obama's decision to remove offshore areas for future leasing that would have allowed exploration and drilling in the Arctic and Atlantic Oceans. The union urged Trump to reverse the decision once he entered the Oval Office.[15] It also launched its Clean Power Progress, an initiative designed to build policy support for gas drilling and exploration. The Laborers' Union avoids using the word *fracking*, but given the fact that every year more U.S. gas comes from fracturing shale rock and less and less comes from conventional gas drilling, the Laborers' position, echoed by NABTU, is to support more fracking in the US.[16]

AN EMERGING CURRENT

As the support of the Trades for the increased development of fossil fuels has become more explicit, so too has union opposition to this same BAU agenda. With the demise of the green growth agenda, this opposition is, in many respects, the starting point for building U.S. trade union support for energy democracy in the years ahead.

Particularly significant and groundbreaking was union opposition to the Keystone XL Pipeline. Among social movements, resistance to the project grew dramatically in 2011 and 2012, led by indigenous organizations, farmer and rancher groups, and environmental, environmental justice, and climate advocates.

Union opposition to KXL, however, emerged slowly. The Laborers' Union support for KXL had been particularly visible and strident. Aware of the likely reaction from the Trades, the Amalgamated Transit Union (ATU) and the Transport Workers Union (TWU) nevertheless chose to be the first U.S. unions to say no to KXL in August 2011.[17] National Nurses United (NNU) and the New York State Nurses Association also issued strong statements, and in mid-2013 more than two thousand nurses and other union members marched over the Golden Gate Bridge (figure 7-1) carrying placards that read "No to the KXL, Stop Climate Change." The National Domestic Workers Alliance, Domestic Workers United, United Electrical Workers, and Local 1199 of the Service Employees International Union (SEIU) would complete the group of unions opposing the pipeline.

It is important to note that the decision on the part of several unions to oppose KXL was anything but pro forma or routine, especially for AFL-CIO affiliates. They knew that the Trades' reaction would be hostile. In a series of statements, Laborers' president Terry O'Sullivan railed against the ATU, TWU, and NNU. Unions opposing KXL, said O'Sullivan, had no right to take food from the mouths of work-deprived union members, and "should come out from under the skirts of delusional environmental groups which [sic] stand in the way of creating good, much needed American jobs."[18]

Other unions chose not to explicitly oppose KXL but were concerned to protect President Obama from Republican efforts to use KXL as a way of hammering him on jobs in a presidential election year. In January 2012,

Figure 7-1. National Nurses United lead march across the Golden Gate Bridge in San Francisco to protest the Keystone XL Pipeline, June 21, 2013. *Image source: National Nurses United*

the Service Employees International Union (SEIU); the United Steelworkers (USW); the Communications Workers of America (CWA); the United Automobile Workers (UAW); and the Retail, Wholesale and Department Union and United Food and Commercial Workers (RWDSU/UFCW) backed Obama's decision to delay a quick yes or no on the pipeline until the environmental and economic impacts had been fully investigated by the State Department.[19] Reacting to the president's decision to delay, the Trades lined up with Republican critics like presidential hopeful Mitt Romney in criticizing President Obama for turning his back on the unemployed and blue-collar America.[20]

It is worth noting that in the midst of all the controversy, BGA refused to take sides in this crucial battle, in the name of preserving the alliance. BGA put considerable time and resources into actively persuading unions like the SEIU and UAW *not* to oppose KXL, remaining publicly silent as the Trades adopted the BAU agenda. Even so, LiUNA, the Plumbers and Pipefitters, and the Sheet Metal Workers left the BGA, accusing unions that had backed Obama's decision of stabbing them in the back.

Standing Rock

Led by Standing Rock Sioux and other native tribes, the struggle over the Dakota Access Pipeline (DAPL) saw unions drawn into another emblematic battle against large fossil fuel interests in 2016. In contrast to the slow accumulation of union opposition to Keystone XL Pipeline, several unions quickly announced their opposition to the DAPL, among them the Amalgamated Transit Union (ATU); the Communications Workers of America (CWA); the National Domestic Workers Alliance (NDWA); National Nurses United (NNU); the New York State Nurses Association (NYSNA); the Service Employees International Union (SEIU); Local 1199 of the SEIU; and the United Electrical, Radio and Machine Workers of America (UE). These unions were also joined by the Labor Coalition for Community Action (LCCA), which represents AFL-CIO constituency groups representing unionists from Latin America (the Labor Council for Latin American Advancement, or LCLAA); the Asia Pacific region (the Asian Pacific American Labor Alliance, or APALA); the LBGT community (Pride at Work, or PAW); African American and Caribbean workers (the Coalition of Black Trade Unionists, or CBTU); women workers (the Coalition of Labor Union Women, or CLUW); and the A. Philip Randolph Institute, an organization for black trade unionists.

Predictably, NABTU and the Laborers criticized both the Standing Rock Sioux and other tribal nations for opposing DAPL and the unions that supported their fight. In an interview, NABTU president Sean McGarvey dismissed the concerns of those who were opposing the fossil fuel projects: "There is no way to satisfy them . . . no way for them to recognize that if we don't want to lose our place in the world as the economic superpower, then we have to have this infrastructure and the ability to responsibly reap the benefits of what God has given this country in its natural resources."[21]

Sharpening Struggle

Reacting to the unions' solidarity with the Standing Rock Sioux, on September 14, 2016, McGarvey wrote a scathing letter to AFL-CIO president Richard Trumka, copies of which were sent to the principal officers of all of the federation's affiliated unions. In a fashion reminiscent of the Keystone XL fight, McGarvey disparaged the unions that opposed DAPL. Referring to the fact that many of the unions that opposed KXL also opposed DAPL, McGarvey wrote: "It seems the same outdated, lowest common denominator group of so-called labor organizations has once again seen fit to demean and call for the termination of thousands of union construction jobs in the Heartland. I fear that this has once again hastened a very real split in the labor movement at a time that, should their ceaseless rhetoric be taken seriously, even they suggest we can least afford it."[22]

A day later, on September 15, the AFL-CIO issued its own already infamous statement supporting DAPL. "Trying to make climate policy by attacking individual construction projects is neither effective nor fair to the workers involved," the statement read. "The AFL-CIO calls on the Obama Administration to allow construction of the Dakota Access Pipeline to continue."[23]

It's probably safe to say that the labor movement is more polarized now than at any time since the early years of the Cold War. Despite the Trades' belligerent support for the BAU fossil fuel agenda, more unions today understand that the extraction, transportation, refining, and disposal of waste by-products associated with fossil fuel use is inflicting intolerable damage on communities and ecosystems, further destabilizing the Earth's climate, and ultimately weakening the U.S. economy and impacting all jobs over the longer term (figure 7-2).

Figure 7-2. Many union delegations were at the forefront of the historic climate march in New York City on September 21, 2014, demonstrating the sharpening struggle within the U.S. labor movement over a climate agenda. *Image source: Al Weinrub*

Toward Energy Democracy

The present struggle between U.S. unions over energy is not unique. Unions in Europe are divided over proposals to begin fracking for shale gas and to expand nuclear power. In Canada, unions in construction and energy-intensive sectors have pushed back against proposals made by other unions to accelerate the energy transition and to create "climate jobs." In Argentina, the main union body, the CGT (Confederación Central del Trabajo), supports fracking, while other unions are opposed. In Germany, unions in coal have often echoed the concerns of the large energy companies with regard to the high costs of renewable power to consumers and businesses.

At the level of global discourse, many unions had embraced the "green growth" narrative as an opportunity to "reboot" social democracy, particularly in the countries of the Organisation for Economic Co-operation and Development (OECD) and especially in Europe, where the European Trade Union Confederation (ETUC) endorsed the EU's Emissions Trading Scheme.[24] In 2007 the United Nations Environment Programme (UNEP) had launched its Green Economy Initiative in partnership with the International Trade Union Confederation (ITUC) and the International Labour Organization (ILO).[25]

However, it is important to note that unions never extended a carte blanche endorsement of "green growth" thinking. They warned that the market had limits, and the public sector remained important. Nevertheless, for unions, "green growth" offered hope in a time of despair, recognition in place of rejection, and a chance to influence proceedings at the global level instead of looking through the window from the cold outdoors. It was, in many respects, a lifeline for a moderate and pragmatic unionism that had more than thirty years earlier rejected militancy and movement building in favor of "social partnership" and "social dialogue."

PROGRAMMATIC SHIFT

The failure of the green growth agenda internationally is therefore an important part of a new reality facing unions. The pursuit of austerity policies and the rise of precarious work in the OECD countries (where most union members reside), along with the ongoing push to privatize services and impose new investment agreements, threaten to further weaken the power of states and governments and have thrown the trade union movement into a crisis of existential proportions.

In common with other social movements, more unions are today willing to question capitalism's ability to meet basic human needs, and they seriously doubt that it can be reorganized in a way that can effectively address the climate crisis in a world economy that is projected to nearly triple in size by 2050.[26]

As a result, a growing number of unions have begun to generate a new trade union discourse on sustainability, climate protection, and the political economy as a whole. The underlying principles are class based, inclusive, and internationalist.

In truth, it is a discourse that owes a considerable debt to other social movements, particularly those connected to organizations in the South representing indigenous people, peasants, and the landless. Unions are now more willing to accept that regulatory and market-based approaches—including carbon pricing—have failed because they do not confront the power of the corporations and their control over energy resources, infrastructure, and markets. These market approaches have also not been able to impede the rush toward rising energy demand, rising fossil fuel use, rising emissions, and higher levels of exploitation of nature, including humans.

An important milestone in the rise of a new trade union narrative was the Second Trade Union Assembly on Labour and the Environment at the Rio+20 talks in Rio de Janeiro in June 2012. Promoting the green growth agenda, UNEP executive director Achim Steiner again pointed to the "inevitable" rise of the low-carbon economy and the equally inevitable creation of millions of green jobs. But after an intense debate among the four hundred delegates, the Assembly declared that the current profit-driven system of production and consumption needs to be replaced, the commons defended, and energy brought into public ownership.[27]

Similarly, also in Rio, the Building Workers' International (BWI) warned that "the current Green Economy concept . . . over-emphasizes market based mechanisms" and could lead to "a green-washing of existing capitalist structures rather than addressing the real causes of the multiple crises."[28] During the same period, the Alliance of Progressive Labor (APL) in the Philippines united with other organizations across Asia in calling for "an immediate stop to the commodification, privatization and financialization of nature, and all its components and functions."[29] Unions like the National Union of Metalworkers of South Africa (NUMSA) and the Canadian Union of Public Employees (CUPE) began to step up their own efforts to bring energy into public ownership and to renationalize what had been privatized.[30]

THE FOCUS ON ENERGY

Building on this momentum, twenty-nine unions, including national trade union centers and global union federations, came together in New York in October 2012 to launch Trade Unions for Energy Democracy (TUED). The meeting discussed the crisis of the green growth agenda and the constraining effect it had had on the search for effective alternatives to market-based ideas.

The three-day meeting in New York was organized around a discussion paper titled *Resist, Reclaim, Restructure: Unions and the Struggle for Energy Democracy.*[31] The paper proposed that the quest for energy democracy entails three broad and strategic objectives: (1) resisting the "business as usual" (BAU) agenda of large energy corporations; (2) reclaiming to the public sphere parts of the energy economy that have been privatized or marketized; and (3) restructuring the global energy system in order to massively scale up renewable and low-carbon energy, aggressively implement energy conservation, ensure job creation and local wealth creation, and assert greater community and democratic control over the energy sector.

The discussion paper also focused on energy poverty. Even though more energy is being generated and consumed with each passing year, more than 1.3 billion people worldwide are without electricity access, and another 1 billion have unreliable access. At least 2.7 billion people lack access to modern, nonpolluting fuels. In many countries, privatization of energy has caused price increases, declining quality and service, and underinvestment.[32]

Since its launch in late 2012, TUED has issued a series of working papers showing that regulatory and market-based approaches to promote renewable energy and energy conservation are totally inadequate given the challenge of climate change and the need to reduce emissions dramatically. Taken together, these papers have confronted the idea that governments could bring about a full-on energy transition by "sending market signals" incentivizing renewable energy. Such an approach fails to deal with the fact that fossil fuel–producing companies and utilities have enormous political and economic power, and are—on the whole— pursuing a BAU agenda that will, if not prevented, eventually destroy the ecosystems that sustain life. A successful energy transition will therefore require a policy shift of major proportions, and it will include bold measures to effectively deal with the wealth, assets, and political leverage of the large energy corporations.

TUED has argued that a timely and equitable energy transition can occur only by way of a successful challenge to existing ownership relations in the energy sector, both in fossil fuels and renewable power. Energy democracy requires that workers, communities, and the public at large have a real voice in decision making, and that the anarchy of liberalized energy markets be replaced with a comprehensive and planned approach.

By the end of 2016, TUED consisted of more than fifty unions (many of them large) from seventeen countries, with several U.S. unions among the most recent recruits. TUED has become an important platform for unions that wish to promote democratic control and social ownership of energy resources, infrastructure, and options.

Key Programmatic Challenges

More unions are using and identifying with the term *energy democracy*, but there exists today only a limited understanding as to what energy democracy might mean in practice. TUED unions understand that to give life to the concept of energy democracy will require rigorous programmatic work. This work is needed to show that a democratic and renewables-based energy system is necessary as one of the main solutions to the climate crisis, energy poverty, and pollution-related illnesses and ecological degradation. Armed with a clear sense of the possible, unions can then search for ways to fight for energy democracy and perhaps engage in the process as intentional economic actors, deploying workers' skills, pensions funds, and other capacities to the energy transition.

What follows is a brief discussion on some of the key issues that will need to be tackled, and hopefully resolved, in the period ahead.

SPEED, SCALE, AND THE ROLE OF THE STATE

For most unions in and around TUED, *energy democracy* refers to a future system that is mostly decentralized, is publicly owned and managed, and is one in which local authorities have an important role to play. But TUED unions also see an active role for the state at the national level. While recognizing that there is little that is intrinsically progressive about state ownership of energy generation and transmission, reclaiming existing energy resources, infrastructure, and options to the public control is considered to be crucially important.

Unions in TUED share the view that achieving energy democracy will entail a wholesale reorientation of most existing public companies, a redefining of the political economy of energy around truly sustainable principles, and a new set of priorities. Some unions, such as the National Union of Metalworkers (NUMSA) in South Africa and CUPE in Canada, have talked in terms of reclaiming or resocializing entities that were once privatized or marketized.[33]

In terms of addressing the climate crisis, the commitments adopted under the Paris Agreement negotiated in late 2015 clearly require decisive government interventions, interventions that go beyond "sending market signals" in the manner imagined by Nicholas Stern and other champions of green growth. Why is this? The Agreement acknowledges the need for global warming to stay "well below 2 degrees Celsius" at a minimum, and it also adopted a "net zero" emissions target by 2060–2070. Net zero will require the full decarbonization of the global economy in just four or five decades. At that point, any greenhouse gases released—to generate electricity; make products; power cars, trucks, ships, and airplanes; heat and cool buildings; raise and slaughter billions of animals; and so forth— must somehow be offset, or "neutralized." In 2015, the global economy generated roughly 57 billion tons of CO_2; almost twice the annual emissions levels of the mid-1990s.[34]

The challenge is for states to fill the huge investment gaps in renewable power in a way that strengthens the public sphere and yields benefits to communities. They must oversee and finance the system upgrades needed to support the introduction of renewables on a large scale. They can and must leverage their access to capital, develop increased research and development, exploit economies of scale, coordinate and provide technical support to local and municipal-level initiatives, and assess the overall status of the energy transition at any given moment. All this should be done in a way that, as far as possible, restricts the levels of involvement and control of large private interests.

MUNICIPALIZATION

Public ownership at the municipal level appears to provide a space in which individuals, communities, unions, and government agencies can open up a "front" in the fight for energy democracy that could bring immediate benefits to cities and communities. According to the Public Services International Research Unit (PSIRU)—the research arm of public sector unions globally—public renewable power generation, combined with the capacity of municipalities to purchase energy, will steadily erode the economic viability of fossil fuel extraction and fossil-dependent utilities without incurring the customary and often debilitating costs of nationalizing or renationalizing fossil fuel resources or generation capacity.

In Germany, the "remunicipalization" of energy has moved forward at a steady pace, and the country now boasts the highest share of renewable energy use in the European Union. In the 1980s and 1990s, a recent report by PSIRU notes how, between 2007 and mid-2012, over 60 new local public utilities (*Stadtwerke*) have been set up, and more than 190 concessions for energy distribution networks—the great majority being electricity distribution networks—have returned to public hands.[35]

In the United Kingdom, the idea of energy democracy entered the national debate when Jeremy Corbyn was elected leader of the Labour Party in September 2015. Corbyn had announced that, if elected, the Labour Party would break up the cartel of private energy companies known as the "Big Six." Corbyn's energy transformation will be driven by "200 green cities," with new municipal energy authorities given a central role in a new set of ownership and governance structures. Three of the largest UK trade unions (UNITE, UNISON, and PCS) have supported Corbyn's commitment.

The United States, as well, has seen effort to "municipalize" energy, principally taking the form of Community Choice energy programs, as described in chapter 8.

Small Is Not Always Beautiful

As is well known, renewable energy systems operate under two main models: centralized and decentralized. *Centralized* systems include such structures as large-scale wind farms and remote central-station solar power plants. The term *decentralized systems* refers to integrated distributed energy resources, such as local renewable generation located on existing buildings or on vacant land close to the point of electricity consumption.

For unions, a decision to support or oppose either the centralized or the decentralized approach is often determined by whether the union concerned is able to represent, protect, and improve the lives of the workers who put these systems in place and maintain them. This is particularly an issue in the residential solar market, where installation companies are small, nonunion contractors who hire via informal networks.

Some unions note that many of those advocating for decentralized generation wish, often in the name of local democracy or energy autonomy, to further liberalize the energy system and undermine the unionized and regulated public utilities. Therefore, the idea of providing grid access and market space to countless numbers of small nonunion installation companies and their supplies is generally not supported by energy sector unions.

Unions are drawn to utility-scale renewable energy projects, because large private energy companies and investor-owned utilities (IOUs) are more likely to hire unionized workers. In contrast, most community-based local energy projects involve contractors that are local and mostly nonunion. This further ties unions to the present, centralized system.[36]

Efforts to build support for the scaling up of renewable power within a democratic framework will need to explore the social benefits and limitations of both centralized and decentralized systems and to assess the feasibility of exerting control over either. But many U.S.-based advocates of energy democracy believe that centralized generation is intrinsically undemocratic. Decentralized generation is, it is sometimes argued, quite the opposite. Few unions would argue with the proposition that decentralized generation is, in theory, far more conducive to local, and therefore democratic, control.

Overall, unions are rarely impressed with "prosumer" approaches to energy democracy that situate individuals, or small groups of individuals, at the center of a new energy vision—essentially "microprivatization." And while unions and worker and consumer cooperatives have close historical ties, unions generally do not regard energy cooperatives to be necessarily progressive or proworker. However, in Australia, the National Union of Workers has developed a proposal whereby its members establish a retail cooperative, with up-front financing coming from the union's general funds.

Although the discussions have barely begun in earnest, unions understand that the role, nature, and perhaps success of energy cooperatives can be shaped by the unions themselves, in the same way as union activists, radical Christians, and socialists pioneered the Mondragon Corporation, a federation of worker cooperatives based in the Basque region of Spain; the Lucas Aerospace workers' Alternative Plan in the United Kingdom; and other important efforts to build cooperatives grounded in a larger social justice agenda.

Another Energy Is Possible

In opposition to both the BAU agenda of the fossil fuel companies and the failed "green growth" agenda of major policy institutions and their business allies, a new agenda is emerging that is grounded in an alternative energy and economic system. The proliferation of energy struggles in the United States draws attention to the fact that these agendas are confronting each other at the level of real politics and real struggles. Nowhere is that more sharply illustrated than in today's trade union politics, where the green growth agenda has been eclipsed by Trump's "America First" energy policy and the vociferous sectionalism of the Trades.

Union opposition to the coal, oil, and gas industry's BAU agenda rose during President Obama's two terms in the White House and will, no doubt, continue under President Trump. This opposition is today in search of its own agenda, one that can inspire union members and working people everywhere to take action. Energy democracy will be an important part of this agenda. Rising resistance will create both the need and the desire to reclaim energy in the United States and elsewhere, and to move decisively toward a renewables-based system that is owned and operated by ordinary people in a manner that can serve everyone without destroying the ecosystems that sustain life.

Democratic control of energy provides a springboard for democratization of the main pillars of the global political economy. Unions are embracing a new, more radical narrative and remain important players in the struggle for a truly sustainable system. But the rigorous programmatic work is still in its early stages and is, in a sense, playing catch-up as political changes open up new opportunities (and new dangers). Another energy is possible. Now, unions and social movements around the world need to make it *a reality*.

1. Table X–2 from "Fortune Global 500" (2013), http://money.cnn.com/magazines/fortune/global500/2013/full_list.
2. Nicolas Stern, *The Economics of Climate Change: The Stern Review* (Cambridge, MA: Cambridge University Press, 2007).
3. As a result of this market-based, green growth influence, the AFL-CIO has, since the 1990s, never supported the greenhouse gas reduction approach of the Kyoto Protocol or the commitment of the International Trade Union Confederation (ITUC) to the science-based emissions reduction targets proposed by the Intergovernmental Panel on Climate Change (IPCC). Every other major national labor body supported these targets and they continue to do so, but not the AFL-CIO. For the ITUC's position, see http://www.ituc-csi.org/IMG/pdf/ituc_contribution_to_unfccc_cop22_en.pdf.
4. European Trade Union Institute, *Exiting from the Crisis: Towards a Model of More Equitable and Sustainable Growth: Report of a Trade Union, Task Force*, edited by David Coats, accessed April 24, 2017, http://www.ituc-csi.org/IMG/pdf/Exiting_from_the_crisis_Washington.pdf; Anabella Rosemberg, "Rio 20: Trade Unions Submit Their Proposals for the Summit," International Trade Union Confederation; Building Workers' Power, ITUC, October 27, 2011, accessed April 24, 2017, http://www.ituc-csi.org/rio-20-trade-unions-submit-their.html?lang=en; Sharan Burrow, "No Social Justice Without Environmental Protection," International Trade Union Confederation; Building Workers, ITUC, June 22, 2012, accessed April 24, 2017, http://www.ituc-csi.org/no-social-justice-without.html.
5. ITUC, "COP15 Evaluation & Report on Trade Union Activities," circulated via e-mail to delegates, February 15, 2010.
6. Robert Baugh, memo to Richard L. Trumka, "Report to the AFL-CIO on the Copenhagen Climate Talks" (unpublished).
7. Intergovernmental Panel on Climate Change, "IPCC Fifth Assessment Synthesis Report," accessed November 29, 2014, http://www.ipcc.ch/pdf/assessment-report/ar5/syr/SYR_AR5_LONGERREPORT.pdf.
8. Oil and Natural Gas Industry Labor-Management Committee, accessed April 24, 2017, http://www.ongil-mc.org/about; Leadership Chairman: Sean McGarvey, president, Building-Construction Trades Department, AFL-CIO; Secretary/Treasurer Jack Gerard, president and CEO, American Petroleum Institute.
9. Importantly, the four union presidents counterpoise jobs and the environment—choosing the former while dismissing the latter. The letter acknowledges the fact that

"further development of Canada's oil sands puts in jeopardy U.S. efforts aimed at capping carbon emissions and greenhouse gases" and that "comprehensive energy and environmental policy should strive to address climate concerns while simultaneously ensuring adequate supplies of reliable energy and promoting energy independence and national security." Union leaders letter to Secretary of State Hillary R. Clinton, unpublished.

10. "Labor Agreement for Keystone XL Pipeline to Create 13,000 American Jobs," *Pipeline News*, posted October 1, 2010, accessed April 24, 2017, https://pipeline-news.com/labor -agreement-keystone-xl-pipeline-create-13000-american-jobs.

11. Koch Industries, Inc., accessed April 24, 2017, http://www.kochfacts.com/kf/americans -for-prosperity.

12. LiUNA and Operating Engineers, Letter to the Honorable John Boehner, Speaker, and the Honorable Nancy Pelosi, Minority Leader, September 9, 2015, accessed April 24, 2017, https://energycommerce.house.gov/news-center/letters/letters-support-hr-702-adapt -changing-crude-oil-market-conditions.

13. The Business Journals, "Union Builds Bridges with Business—and Even Some Republicans" (video), posted April 22, 2015, accessed April 24, 2017, http://www.bizjournals.com /bizjournals/washingtonbureau/2015/04/union-builds-bridges-with-business-and-even -some.html. The full quote: "Even if you look at Koch Industries—they're one of our biggest clients. You'll never see us making public statements saying negative things about Koch Industries. They're a huge client of ours. Do we agree with some of the things that they supposedly support? No. Do we understand why they do it? Yeah, OK, because they're looking for political advantage for a political point of view, and the Democrats don't see it the way they see it. And other unions in the labor movement tend to be much more Democratic unions. And if you can hurt the labor movement, i.e., you hurt the Democratic Party. It's just a system that we really don't want to be engaged or involved in."

14. North American Building Trades Unions (NABTU), press release, posted December 13, 2016, accessed April 24, 2017, https://nabtu.org/press_releases/nomination-rex-tillerson -u-s-secretary-state.

15. LiUNA press statement, "Ban on Offshore Drilling—Another Bad Energy Proposal," posted December 22, 2016, accessed April 24, 2017, http://www.liuna.org/news/story /ban-on-offshore-drilling%E2%80%94another-bad-energy-proposal.

16. LiUNA, "Clean Power Progress," accessed April 24, 2017, http://cleanpowerprogress.org /wp-content/uploads/2016/07/Factsheet-UnitedStates-11.pdf.

17. ATU and TWU, "ATU & TWU Oppose Approval of the Keystone XL Pipeline and Call for End of Increased Use of Tar Sands Oil," press release, August 19, 2011, accessed April 24, 2017, https://www.atu.org/media/releases/atu-twu-oppose-approval-of-the-keystone-xl -pipeline-and-call-for-end-of-increased-use-of-tar-sands-oil.

18. Statement, LiUNA, September 9, 2011. On November 11, 2011, O'Sullivan wrote in Nebraska's *Journal Star*: "LIUNA members are angry and surprised that some of our fellow union brothers and sisters, who don't have any members in the pipeline industry and who themselves have never worked on a pipeline, now speak out as pipeline experts in their attempt to block Keystone. For thousands of LIUNA workers, the stakes are too high to not hold them accountable; LiUNA President Terry O'Sullivan, "Keystone XL opponents have let down America's workers," accessed April 24, 2017, http://journalstar.com/news /opinion/editorial/columnists/local-view-keystone-xl-opponents-have-let-down-america-s /article_a001d329-2869-5275-acd1-5347f79cadf6.html.

19. Communication Workers of America (CWA), "Environmental Groups, Unions Support President's Decision on Keystone XL," posted Wednesday, January 18, 2012, accessed April 24, 2017, http://www.cwa- union.org/news/entry/environmental_groups_unions_support _presidents_decision_on_keystone_xl#.UPRjyrZgPrE.

20. Building and Construction Trades Department, AFL-CIO, Mark Ayers (president), http://www.christianpost.com/news/obamas-keystone-decision-alienates-him-from-the -majority-of-americans-67553/#0rqCqD9thhzAh08q.99.

21. "Natural Gas Now, the Next Infrastructure Challenge: Conversation with Sean McGarvey" (video), posted May 14, 2015, accessed April 24, 2017, emberhttps://youtube/r89lKwockYY.

22. Sean McGarvey NABTU, letter to Richard Trumka, AFL-CIO, September 15, 2016.

23. AFL-CIO, press statement, "Dakota Access Pipeline Provides High Quality Jobs," September 16, 2016, accessed April 24, 2017, http://www.aflcio.org/Press-Room/Press-Releases /Dakota-Access-Pipeline-Provides-High-Quality-Jobs.

24. "The European Social Model is an example for the rest of the world of a society based on social justice and solidarity, where economic and social advancement take equal priority, and where decent work and social protection combat poverty and social exclusion. That is why the success of Social Europe is so important not only for European citizens, but also for developing just and fair political systems in other countries." ETUC, June 2006; ETUC, Brussels, June 5, 2007, ETUC position on the revision of the EU Emissions trading directive, https://www.etuc.org/documents/etuc-position-revision-eu-emissions-trading-directive#. VomRKFKGbX8; Trade Union Sustainable Development Advisory Committee (TUSDAC), 2007, cited by Paul Hampton, *Workers & Unions for Climate Solidarity: Tackling Climate Change in a Neoliberal World* (Routledge, June 16, 2015), chapter 4; ETUC, ETUC's position on the climate change and energy package, position adopted by ETUC Executive Committee, March 4, 2008 (Brussels), p. 1, https://www.etuc.org/documents/etuc%E2%80%99s -position-climate-change-and-energy-package#.VpKnUlKGbX8.

25. International Trade Union Confederation, "ITUC Growing Green and Decent Jobs," April 2012, http://www.ituc-csi.org/IMG/pdf/ituc_green_jobs_summary_en_final.pdf.

26. PriceWaterhouseCoopers (PwC), *The World in 2050*, February 2015. PwC analysis is based on data from the World Bank, IMF, and other sources.

27. "Trade Union Resolution on Labour and Environment," ITUC, June 11–13, 2012, posted January 16, 2014, http://www.sustainlabour.org/documentos/Trade Union Resolution on Labour and Environment.pdf.

28. BWI Conference on Sustainable Industries, "Fighting Back for Sustainable Development in Construction and Forestry," Declaration on Rio+20 Summit, June 11–14, 2012, http://www.ituc-csi.org/IMG/pdf/bwi_rio_declaration_2012_-_final.pdf.

29. Alliance of Progressive Labor (Philippines), "Fight for Our Future! No Price on Nature!," Statement dated June 23, 2012, accessed April 24, 2017, http://www.apl.org.ph/?s=nature.

30. Canadian Union of Public Employees, *Working Harmoniously on the Earth*, CUPE's National Environment Policy, CUPE.SCFP, accessed January 26, 2014, http://cupe.ca/updir/Working _harmoniously_on_the_Earth_-_FINAL.pdf. According to CUPE "We will work to keep energy generation and transmission public and promote public renewable energy, including advocating for bringing energy generation and transmission back into public ownership and control where it has been privatized." See also Statement by National Union of Metal Workers of South Africa (NUMSA), December 2012, NUMSA central committee meeting (CC), held December 11–14, at Vincent Mabuyakhulu Conference, "We believe

a just transition must be based in worker-controlled, democratic social ownership of key means of production and means of subsistence. . . . Without this struggle over ownership, and the struggle for a socially owned renewable energy sector, a just transition will become a capitalist concept, building up a capitalist 'green economy.'"

31. Sean Sweeney, "Resist, Reclaim, Restructure: Unions and the Struggle for Energy Democracy," 2012, Trade Unions For Energy Democracy, accessed April 24, 2017, http://unionsforenergydemocracy.org/resources/tued-publications.
32. International Renewable Energy Agency (IRENA), *Renewable Energy Jobs & Access* (Abu Dhabi, 2012).
33. National Union of Metalworkers of South Africa, statement from International Conference on Building a Renewable Energy Sector in South Africa, Johannesburg, February 4–8, 2012.
34. Global Carbon Project, *Global Carbon Budget, 2015,* accessed April 24, 2017, http://www.globalcarbonproject.org/carbonbudget/15/files/GCP_budget_2015_v1.02.pdf.
35. Public Services International Research Unit, "Energy Liberalisation, Privatisation and Public Ownership," September 2013, David Hall, Sandra van Niekerk, Jenny Nguyen, Steve Thomas, accessed April 24, 2017, www.world-psi.org/sites/default/files/en_psiru_ppp_final_lux.pdf.
36. Ibid.

CHAPTER 8

Democratizing Municipal-Scale Power

AL WEINRUB

On June 21, 2016, Pacific Gas and Electric (PG&E), the largest electrical utility in California and one of the largest in the country, announced that it would be shuttering its Diablo Canyon nuclear power plant, the last remaining nuclear facility in the state.

In announcing its decision to forego operation of the plant beyond its current 2025 license, PG&E cited a number of factors that capture the rapidly changing electric power landscape in the state and across the country, including new developments that "will significantly reduce the need for Diablo Canyon's electricity output." These included state mandates for renewable energy and energy efficiency and the growth of distributed energy resources. But PG&E also cited "potential increases in the departure of PG&E's retail load customers to Community Choice Aggregation."[1]

Hidden in that brief mention is a pitched battle that has been taking place in California over recent years between advocates of community-controlled renewable energy systems and the state's three investor-owned monopoly utilities. In announcing the 2025 closure of Diablo Canyon, PG&E essentially admitted that it is losing that battle, acknowledging that because of the expected loss of customers to Community Choice energy programs in California, there would be insufficient demand in the future for PG&E's nuclear power.

So what is Community Choice energy, and what is its potential for establishing a new energy model that can democratize energy, both in California and in other states?

A Public Energy Services Provider

For more than a century, electricity supply has been virtually a monopoly enterprise, and consumers have had little say in how their electricity was procured and delivered. That has now changed for energy consumers in California and a few other states.

Community Choice Aggregation, as it is sometimes called, is a mechanism that allows cities, counties, and a few other government entities to aggregate individual electricity customers within a defined service area for the purpose of providing electricity and related energy services. Six states besides California (Illinois, Ohio, Massachusetts, Rhode Island, New Jersey, and most recently, New York) allow local governments to procure their own electricity supplies in this way, while the incumbent utility continues to operate the electricity transmission and distribution infrastructure. Community Choice programs are "opt-out" initiatives, meaning that the program can automatically enroll electricity customers in their jurisdiction, but those customers can choose to opt out and stay with the incumbent utility at any time.

Community Choice programs provide local control over energy supply (but are distinct from municipal utilities, which also own the distribution infrastructure). However, these programs are not limited to buying and selling electricity. They are also about managing a community's energy resources (both for reducing electricity demand and for generating electricity) to meet local objectives. Woody Hastings of the Center for Climate Protection in Sonoma County, California, one of the jurisdictions that has opted for a Community Choice energy program, puts it this way:

> Community Choice puts our community in control of the most important part of our electricity system. That means we can purchase more renewable and greenhouse-gas-free energy on the market than PG&E offered us. But more importantly, we can build

renewable energy assets right here in the County. We not only get the benefits of low carbon electricity, but we get the economic benefits—the business opportunities and clean energy jobs—that come from investing in our own community.[2]

Sonoma County's Community Choice customers get power that is 30% lower in greenhouse gases than that of PG&E. They also pay up to 9% less on average than PG&E customers. In addition, electricity net revenues go back into the community rather than into the pockets of PG&E shareholders and highly paid executives.

Sonoma County is one of five communities that, as of this writing (May 2017), have established Community Choice programs in California, and it is estimated that as many as 60 percent of utility customers could depart to Community Choice programs during the next five years.[3]

Based on local preferences for increased levels of renewable energy sources, Community Choice programs can spur investment in local energy resource development, reduce greenhouse gases, procure electricity at lower prices, enhance community resilience. and provide the impetus to modernize the electricity grid.

To be more explicit, Community Choice programs in California offer a number of potential benefits for local communities:

- **Local control**. Community Choice gives communities control over where their electricity comes from and how their electricity dollars are spent. A Community Choice agency would be governed by a public board of directors. Through this public governance structure, communities can have a say in the program's goals, how it operates, and the types of resources it procures. Surplus revenues can be leveraged to stimulate development in the community: investments can be made in demand reduction (such as energy efficiency), renewable energy development, energy storage, and so forth. In this way, local citizens can participate in shaping the program to address community needs.
- **Local choice**. Community Choice programs are essentially about giving consumers the choice of an alternative electricity service provider

they would not otherwise have. Under the current investor-owned utility model, most consumers can buy power from only one company, with no say about where that power comes from or how the revenues are used. This means that consumers unhappy with the utility have nowhere to turn except to a state's regulatory body, such as the California Public Utility Commission (CPUC), which generally serves utility interests, not ratepayer interests.

- **Local economic development benefits**. Community Choice programs have the ability to develop demand reduction and storage resources, as well as solar, wind, and other renewable resources in or near their service areas. Unlike traditional utility electricity sources that are remote from communities (and send power over long-distance transmission lines), locally developed resources represent investment in the local economy. This investment can create meaningful economic benefits, including growth in clean energy jobs.

  Because Community Choice agencies can generally finance projects with tax-exempt revenue bonds—which incur lower financing costs than private financing—and do not have to pay dividends to shareholders, more net revenues from a local development program would stay within the local community.[4] The community can decide how these proceeds are utilized (for example, to lower electricity rates, or to create new incentive programs, or to build a contingency reserve).

  Finally, local economic benefits accrue also to local property owners and businesses from energy savings and on-site electricity generation that can be encouraged by the Community Choice program. For many commercial building owners, renewable energy development can mean increased revenues, both from direct investment and by offering leasing rights to project developers.

- **Environmental benefits**. By reducing demand and procuring more electricity from renewable resources, the Community Choice program can substantially reduce greenhouse gas (GHG) emissions associated with electricity consumption. As many municipal climate action plans have indicated, a major source of GHG emissions is from electricity generated by fossil fuel combustion in power plants.[5]

- **New local energy programs**. A Community Choice agency can develop programs for demand reduction and new renewable generation that are very difficult to achieve at the state level. For example, the agency can promote energy efficiency and demand response programs above and beyond what the incumbent utility offers. These programs can be designed specifically to meet the needs of the community.

  In addition, the Community Choice program can incentivize local renewable electricity generation through well-designed net metering,[6] feed-in tariffs,[7] and shared renewables[8] programs, as well as other ways of aggregating, collectivizing, and financing new energy resources.

- **Rate stability and lower prices**. By focusing on demand reduction and the deployment of renewable resources, a Community Choice program offers the advantage of greater rate stability. Local assets reduce the risks of a volatile energy market. In addition, the continued decline in renewable power generating costs can translate into *lower* rates over the long term.

These benefits reflect the vision of Community Choice as a municipal-scale, public electricity services provider responsive to the economic, environmental, and equity needs of communities. It is an alternative in which community-controlled and community-owned electricity services become an important expression of energy democracy and pave the way for the creation of a new renewable energy model.

A New Renewable Energy Model

As indicated above, some of the main potential benefits of a Community Choice program derive from the development of community-based renewable energy resources. This new model of energy development—the decentralized renewable energy model—stands in contrast to the legacy model of fossil fuel electrical energy production, the centralized energy model of coal, natural gas, and nuclear power plants.

The centralized energy model, even when applied to *renewable* energy, is based on large-scale, centralized generating systems—big solar plantations and large wind farms—that are the product of concentrated financial and economic power. In most cases, centralized energy development represents the interests of powerful economic forces aligned with investor-owned utilities and aided by a corporate/state apparatus unfettered by democratic restraints.[9]

By contrast, the decentralized renewable energy model supports community-based renewable energy development. It fosters the economically sustainable, ecologically sound, and equitable relationships needed by communities to address the current economic and climate crisis.

Thus, the decentralized renewable energy model enables not only the shift from fossil fuel power to renewable power, but also the shift from corporate control of energy systems to more democratically controlled energy systems. This provides a basis for community-based *decentralized* development of distributed energy resources (such as solar energy, wind, geothermal energy, energy conservation, energy efficiency, energy storage, and demand response systems) at the local level through popular initiatives.

Physically speaking, decentralized energy systems consist of three kinds of distributed energy resource (often referred to as DER) components: *decentralized electricity generation, demand reduction,* and *system balancing,* as illustrated in figure 8-1.

- **Decentralized generation.** This component refers to renewable electricity generation, usually smaller scale, located on existing structures or vacant or contaminated land close to the point of electricity consumption, so that the high cost and energy loss of high-voltage transmission lines is not required. The renewable energy source can be whatever is naturally available in the geographical region, for example, solar, wind, geothermal, small hydro, combined heat and power, or biomass/biogas.

    In the case of solar photovoltaic generation, for example, the energy source can consist of installations on rooftops, parking lots, brownfields, rail or highway rights-of-way, and so forth. It might be

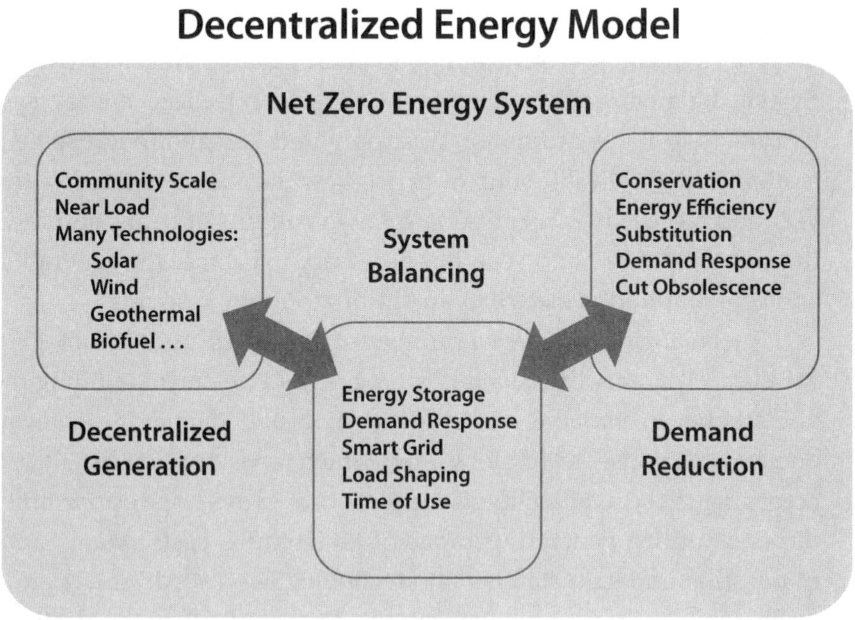

Decentralized Energy Model

Figure 8-1. Graphical representation of a decentralized energy system *Image source: LCEA*

as small as a system of a few kilowatts on a residential building, a 1 megawatt system on a large commercial building (like a large box store), or a ground-mounted 5 megawatt or larger system built on degraded industrial land.

- **Demand reduction**. This component refers to the many technologies for reducing the consumption of electricity. It might include, for example, conservation (turning off the lights); energy efficiency (more-efficient lightbulbs); substitution (use of natural light when possible); demand response (not everyone turns on the lights at the same time); and simply eliminating built-in obsolescence or other forms of waste that consume electricity (one good light that lasts as long as ten poor ones).

  Demand reduction is perhaps the most important component of a decentralized energy system. The cheapest electricity is the electricity that is never produced. For example, the cost over time of retrofitting buildings to conserve energy can be much less than the cost of generat-

145

ing the equivalent amount of electricity. Reducing electricity consumption is also the most ecological way to phase out fossil fuel electricity.

- **System balancing**. This component of a decentralized energy system refers to the coordination of supply and demand. Because generation from renewable sources varies over the course of the day and year and electricity consumption follows patterns of peaks and lows throughout the day and year, it is necessary to balance the generation and consumption of electricity to optimize energy resources.

The balancing involves a number of strategies and technologies. Increasingly competitive battery storage is a key element in filling the voids between variable energy generation and demand. However, demand response technology—by which consumption is altered according to the availability of supply—provides a huge opportunity to better utilize generating capacity and reduce costs. Many communication and grid stabilization technologies—called "smart grid" technologies—are under development. These will allow the electrical distribution system to support demand response technologies and provide the bidirectional flow of electricity and information needed for balancing a decentralized energy system.

Decentralized energy systems are designed to utilize local energy resources—both demand reduction and new generation, along with energy storage and smart system balancing—to meet the electricity needs of their host communities.[10] While this approach requires a great deal of new investment to achieve net-zero energy targets (that is, the community generates what it consumes), the investment can be readily paid off through overall system savings and energy independence.

The decentralized renewable energy model provides a powerful alternative to the traditional nondistributed, nonintegrated centralized energy model. The decentralized model is one that is more ecologically sound, more economically beneficial to communities, more effective in creating local employment, more sustainable, and more *open to community participation in the control of its energy resources*.[11] Community Choice energy programs are a natural vehicle for implementing this new energy model.

More Than Just Another Utility

The above discussion has tried to make the case that Community Choice energy, by placing control of the electricity system in community hands, provides a vehicle for creating the kind of decentralized energy system that can deliver a host of economic, environmental, and equity benefits to our communities.

That is not to say that such benefits are a foregone conclusion.

In fact, many Community Choice energy programs have led to quite different results. Take the case of Illinois, for example, where a few years ago hundreds of communities established Community Choice programs and on that basis were able to shift their purchase of electricity from Consolidated Edison, which had procured relatively costly coal-based electricity sources, to new electricity providers based on cheaper fracked natural gas electricity-generating sources. That meant cheaper electricity for those communities; it also meant an expansion of the extreme fossil fuel extraction method called fracking.

Many of these communities, in an effort to claim that they were reducing greenhouse gas emissions, purchased large numbers of unbundled renewable energy certificates (RECs) on what is called the voluntary REC market. Basically, these unbundled RECs are simply paper certificates and do not add to new renewable energy production.[12] In fact, in Illinois, because these RECs were used to greenwash fossil fuel–sourced energy, the market for real renewable power nearly evaporated, suppressing wind energy production in the state.

So the impact of Community Choice in Illinois was not only to encourage fracking but to suppress wind-powered renewable energy in the region as well.

The lesson of this story is that Community Choice is merely a vehicle; it is not a destination. Without a clear destination and a good driver, this vehicle can take us in the wrong direction, to the wrong place.

For a Community Choice program to deliver economic, environmental, and equity benefits to our communities, it cannot be seen as just another locally based utility that simply buys and sells electricity to residents and businesses. Nevertheless, a number of Community Choice

programs in California, like those in Illinois, are based primarily on purchasing electricity on the market or from remote generating sources for
sale to their customers. This approach is known as Community Choice
Version 1.0.

To achieve the kind of decentralized energy system that can deliver
economic, environmental, and equity benefits to our communities
requires a different community-development approach, known as Community Choice Version 2.0.[13]

Community Choice Version 2.0 is substantially different from the
standard utility model, as shown in table 8-1.

The Strategy: Put the Community in Community Choice

The powerful potential of Community Choice Version 2.0 energy programs
to deliver economic, environmental, and equity benefits to our communities
rests with our communities exercising real control of these programs. While
Community Choice represents a shift of energy decision making away from
the incumbent private utility into a public agency, such institutional restructuring will represent a democratization of energy *only if* our communities are
actively involved in shaping Community Choice programs.

Hence, the basic strategy is to build a public constituency strong
enough to achieve the establishment of a Community Choice Version 2.0
program and hold those governing that program accountable to serving
community needs (figure 8-2).

For that to happen, the program must provide real value to the community. It must be an economic development platform that can build
community wealth—business opportunities and jobs—and do so in a
way that reverses historical patterns of discrimination, all while addressing the impacts of climate change.

But that kind of program requires building a political base in those
communities that would benefit most from such a program—a base cen-
tered in working-class communities, low-income communities, and communities of color. It requires a political constituency strong enough to
shape the electricity system and to make renewable energy a resource for
empowering local communities.

Table 8-1—Comparison of Community Choice Version 2.0 with Investor-Owned Utility Model

Community Choice Version 2.0	Investor-Owned Utility
Nonprofit public agency	For-profit private corporation
Purpose is to maximize community benefits: GHG reduction, economic development, good clean energy jobs, rate stability, social equity, local ownership and control of energy, and other community benefit goals	Purpose is, by law, to maximize shareholder returns
Net electricity revenues remain in the community to expand services, invest in new assets, build reserves, or reduce rates	Net electricity revenues leave the community as utility profits and shareholder dividends
Based on an energy services provider model: provides optimum energy services to community: cuts waste, reduces demand, lowers overall system costs of electricity service	Based on a utility model: buys and sells electricity to ratepayers; the more electricity delivered, the better*
Implements a decentralized renewable energy model: local distributed energy resources are developed to optimize the electricity system, provide stability, and achieve net-zero energy	Implements a centralized renewable energy model: emphasis is on expanding infrastructure investment
Encourages strong community participation in shaping the program and in governance	Decisions made by utility executives and state regulatory bodies serving the utilities

*The standard regulated investor-owned utility model does not allow the utility to profit directly from electricity sales, only from the delivery of electricity, based on a guaranteed return on investments in distribution and transmission infrastructure.

That political constituency is a broad cross-class alliance but is led by those sectors that have the most stake in social justice, equity, resilient communities, and life-sustaining economies.

PRINCIPLES OF DEMOCRATIZED ENERGY DEVELOPMENT

A number of principles express how local energy development can advance democracy and promote the empowerment of working-class communities, low-income communities, and communities of color. Devel-

Figure 8-2. Representatives of community organizations participated in a December 15, 2015, gathering to build support for a Community Choice 2.0 program in the East Bay. *Image source: LCEA*

opment of community-based decentralized energy under a Community Choice energy program can be guided by the following broad principles:

- **Social justice and equity**. Making sure that local energy resources benefit working-class communities, low-income communities, Indigenous communities, and communities of color.
- **Energy democracy**. Enabling community ownership and control of energy resources, with shared leadership and decision-making authority that involves all stakeholder communities.
- **Clean energy jobs and family-sustaining livelihoods**. Creating local jobs, new businesses, and new ownership opportunities that help improve the environment and restore the economies of our communities.
- **Workforce development**. Committing to workforce development programs that create family-sustaining jobs for local residents, especially for those historically disadvantaged and most vulnerable to poverty and pollution.

- **Sustainability.** Respecting ecological interdependence and the limited restorative capacity of the biosphere, while creating the environmental conditions needed to support present and future generations.
- **Healthy communities.** Supporting locally resilient, healthy food systems; affordable, reliable, and accessible public transportation; clean air; clean water; and safe, efficient, affordable housing.
- **Community resilience.** Strengthening vulnerable communities to withstand the impacts of climate change, including disaster scenarios.
- **Social safety net**. Making special provisions for those people unable to afford energy services at normal rates, providing energy security.
- **Precaution**. Accepting that a project, policy, or decision should not be pursued if its impact on human or environmental health is risky or unknown.

On the basis of these principles, communities can advocate for Community Choice energy programs that contribute to vibrant and equitable regional economies. For example, Appendix A at the end of this chapter shows the specific Community Choice energy program goals put forward by San Francisco's East Bay Clean Power Alliance reflecting these principles.

DESIGN AND PLANNING OF THE COMMUNITY CHOICE PROGRAM

To be able to deliver economic, environmental, and equity benefits to our communities, as described earlier, a Community Choice program needs to serve as a platform for developing a decentralized renewable energy system. That means building community-based demand reduction and new generation assets and optimizing the system through energy storage, load shaping, and demand response technologies. The basic features of such a Community Choice Version 2.0 program are illustrated in figure 8-3.

In this concept illustration, the community is engaged in setting goals that inform a business plan for the build-out of local renewable energy resources. That plan is based upon the building of local demand reduction assets and local renewable generation assets over time, and integrating those assets with one another and with the decreasing amounts of energy procured on the market as local assets are built out.

The development and integration of these energy resources is then implemented through a number of initiatives shown at the bottom of figure 8-3:

- Developing customer-side (behind-the-meter) resources.[14]
- Encouraging community-scale development through incentive programs like feed-in tariffs, shared renewable facilities, and energy cooperatives development, as well as through financing programs like Property Assessed Clean Energy (PACE)[15] and on-bill repayment.[16]
- Establishing workforce development programs and labor standards to prepare community members for the new clean energy jobs being created.
- Encouraging technological innovation and new business development.

Appendix B shows the impact on the community of the features described above.

The program design concept of figure 8-3 is meant to facilitate feedback between the various levels shown in the illustration, so that the community is involved in shaping the Community Choice program to address environmental and economic justice. Community engagement in the shaping and governance of the program represents the democratizing of energy made possible through Community Choice.

CONTINUOUS ENGAGEMENT OF THE COMMUNITY

As mentioned earlier, building a strong political constituency is key to establishing a Community Choice energy program of the type being described.

However, it is equally important that the community be engaged in an ongoing basis over the lifetime of such a program (similar to the kind of engagement shown in figure 8-4) We have seen many examples of public (municipal) utilities that have acted no differently from investor-owned utilities, and of rural electric cooperatives—such as those formed under the rural electrification program during the 1930s—that have

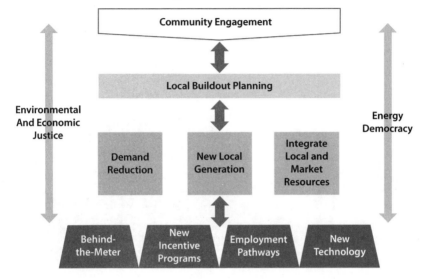

Community-Benefit-Focused Program Features

Figure 8-3. Features of a Community Choice program focused on community benefits
Image source: LCEA

developed ossified governing boards that have lost all touch with the needs of their members.

For a Community Choice energy program to meet community needs, community stakeholder engagement needs to be institutionalized in the program. This implies a governance structure that represents the diverse interests of the community. It also implies strong involvement, both directly and through elected representation, of the community in decision making regarding the design, implementation, and operations of the Community Choice program.

Often, the governing board of a Community Choice program will create community advisory committees to provide input to decision makers. More significant would be direct representation from community stakeholder interests on the governing board itself, perhaps as ex officio (nonvoting) members in cases in which the law requires that only elected officials serve on the board.

Figure 8-4. Workshop session on Community Choice to engage the East Oakland community, April 2014 *Image source: LCEA*

Challenges to Democratizing Municipal-Scale Power

The vision of Community Choice being described in this chapter is not easy to implement. There are challenges at every turn.

Take the experience of those advocating for Community Choice Version 2.0 in California as an example.

The first challenge is opposition from the legacy centralized energy system—the institutions and the ideology that support it. The investor-owned utilities are fighting hard to maintain their monopoly control, with the willing assistance of the state regulatory bodies that support these utilities.

While the Community Choice Aggregation law, Assembly Bill (AB) 117, was passed in 2002, the first Community Choice program in the state, in Marin County, was not established until eight years later, due in large part to a blistering attack by PG&E, the local investor-owned utility, to derail these efforts.

In 2010, PG&E launched a state ballot initiative that would have changed the California Constitution in such a way as to make Community Choice programs all but impossible to establish. Despite PG&E's spending about $50 million on that effort—outspending the grassroots opposition by 5,000 to one—the ballot initiative was defeated. This David-versus-Goliath victory gave a great boost to Community Choice advocates.

Not to be deterred, however, Goliath struck again in 2014, this time with a legislative effort—AB 2145—also designed to kill Community Choice in California. PG&E's union, the International Brotherhood of Electrical Workers (IBEW) Local 1245, and the California Labor Federation led the charge in supporting the monopoly utilities.

Again, grassroots advocates fought back, emphasizing the community and jobs benefits possible under Community Choice. They formed a "No on AB 2145" campaign supported by two hundred organizations across the state, including environmental justice organizations, local governments, and rooftop solar businesses (see the campaign poster, figure 8-5). The campaign created such a storm of protest that the bill, after being passed in the California Assembly, never got introduced on the floor of the California Senate. This victory gave another big boost to Community Choice initiatives throughout the state.

There are now more than seventy jurisdictions in the state, individually or collectively, investigating or

Figure 8-5. Poster used to oppose the monopoly utilities' legislative effort to kill Community Choice in California, July 2014 *Image source: Dignidad Rebelde: Melanie Cervantes and Beto Fuentes*

establishing Community Choice programs. Six programs are currently up and running and a larger number are expecting to launch in 2017, despite determined efforts by the monopoly utilities to crush this emerging movement.

ENTER THE CALIFORNIA PUBLIC UTILITIES COMMISSION (CPUC)

The utilities are aided in their opposition to Community Choice by the CPUC, the state agency set up to regulate the monopoly utilities.

Michael Picker, president of the CPUC, reflecting the regulating body's stance, has expressed open hostility to Community Choice, using the pejorative term *forced collectivization* to characterize the Community Choice alternative to investor-owned utilities.[17] Picker's view is that the prime role of the CPUC is to defend, protect, and perpetuate investor-owned monopoly utilities in the state:

And the question is, where do we need to maintain that monopoly? That's what my agency does. We award monopolies where there's not a market and then we protect them against ruinous or calamitous competition. That's the language that's embedded in our bone and in our blood from the 1910s. There was a thought that that was the best way to mobilize capital—you created a monopoly and you enforced it.[18]

In a few words, Picker explicitly states what critics of the CPUC have long alleged—that the CPUC is a captive agency, serving monopoly utility shareholders and putting investor-owned utility interests ahead of the California public's interest in community-based clean power development.

Picker's views have been manifested in actions detrimental to Community Choice. In 2016, the CPUC doubled the Power Charge Indifference Adjustment (PCIA) awarded to PG&E. The PCIA is an ongoing fee assessed on departing Community Choice customers to compensate the utility for stranded contract costs, and the fee will increase again in 2017, making it more difficult for Community Choice programs in PG&E's service territory to compete with the utility.

And on August 18, 2016, the CPUC gave the green light to San Diego Gas and Electric (SDG&E) to market against Community Choice energy programs in Southern California, a practice prohibited by the 2002 Community Choice law.

Furthermore, as Community Choice has gained momentum, the CPUC has approved monopoly utility proposals to shift billions of dollars in costs, in order to artificially lower utility electricity charges on customers' bills. For example, in PG&E's 2011 General Rate Case, the CPUC approved $3 billion in cost shifting from the generation portion of the electricity bill to the transmission side of the ledger in order to make PG&E's electricity charges relatively lower compared to Community Choice electricity charges.

Such actions by the CPUC, and those under consideration in 2017, undermine the viability of Community Choice energy programs in California.

Other state regulatory agencies have also been problematic. The California Independent System Operator (CAISO), for example, has failed to correct a market distortion by which all electricity customers pay for high-voltage transmission infrastructure, even when their electricity is not delivered through the transmission system. This means the avoided costs of local electricity generation are denied to ratepayers, and communities lose one of the key benefits of local energy development. This distortion makes it more difficult for Community Choice programs to prioritize the development of local renewable energy resources.

LOCAL CHALLENGES

A decentralized, community-based energy system and the possibility of a community-controlled public electricity services agency are hard for many to imagine after many years of a monopoly-dominated electricity system. Electricity is seen as a commodity by most people, not as a basic resource for meeting community and human needs. Hence, there are strong conceptual barriers for communities to overcome when advocating for a democratized energy system.

In addition, local governments and public agencies tend to be very cautious, opting in most cases for the easiest and quickest approach to Community Choice—Verson 1.0: simply purchasing renewable energy at most favorable prices on the market. A Community Choice program that focuses on prioritizing local renewable resource development—Version 2.0—requires upfront planning and design and a strong commitment to community benefit goals. It takes a kind of vision and courage not often found among local politicians.

Version 2.0 also requires persistent organized advocacy from community members (figure 8-6). For low-income communities and communities of color, affordable housing and displacement, police brutality and neighborhood safety, pollution and hunger, and a host of other immediate issues take priority over energy concerns and long-term economic development.

These factors make it hard to build the strong political base needed to democratize energy.

Beyond these political difficulties are the challenges of financial markets that are vested in large-scale centralized energy development and local banking institutions that are averse to investing in new energy models.

In short, there are many challenges to building a constituency with the vision, resources, technological savvy, and commitment needed to force local governments to establish Community Choice energy programs that can deliver economic, environmental, and equity benefits to our communities. This is especially true in the face of strong opposition from the state's monopoly utilities and their regulatory henchmen.

Accomplishments in California

The five existing Community Choice programs in California, and many of the initiatives to establish new ones, do not yet reflect the transformative vision described in this chapter.

Nevertheless, as noted earlier, two statewide efforts on the part of the investor-owned utilities to kill Community Choice—one a state proposition and the other a state legislative bill—have been defeated against all odds. In the aftermath of these victories, the California Alliance for Community

Figure 8-6. Panelists speaking on the significance of Community Choice energy as a driver for local job creation at an April 5, 2014, event in East Oakland: from left, Dominic Ware (OUR Walmart); Margie Castillano (Youth UpRising and Castlemont High School SUDA); Agustin Cervantes (ILWU Local 6); Jakhiyra McDaniel (Youth UpRising and Castlemont High School SUDAWorks); and Nile Malloy (Communities for a Better Environment) *Image source: LCEA*

Energy was created to promote community-based renewable energy development in the state and defend against attacks like those mentioned. The alliance pulls together most Community Choice advocates across the state who agree with its mission to "support and defend Community Choice energy programs in California that advance local clean energy for the environmental and economic benefit of our communities."[19]

Organizations like the Local Clean Energy Alliance and the California Environmental Justice Alliance have upheld the centrality of equity and community decision making in these efforts, an important counter to the natural tendency of public agencies to become bureaucratized and undemocratic.

In the San Francisco Bay Area, a coalition of Community Choice advocates, the East Bay Clean Power Alliance, has organized to bring a

Community Choice Version 2.0 program to Alameda County. A several-year campaign, based on mobilizing community organizations, has gained traction based on principles of democracy and equity. On October 4, 2016, pressured by East Bay activists, including the Alameda Labor Council (figure 8-7), the Alameda County Board of Supervisors voted to approve establishing a Community Choice energy program with a commitment to maximizing community benefits and including community input in the governance of the program. Supervisors allocated $500,000 for development, within eight months of agency formation, of a business plan for achieving local renewable energy development, union and family-supporting jobs, and other community benefits.

A similar effort in South Bay, Los Angeles, has pulled together twenty

Figure 8-7. Martha Kuhl, first vice president of the Alameda Labor Council and leader of the California Nurses Association, speaks in favor of the joint community/labor Unity Proposal at a press conference held outside the Alameda County Administration Building on October 4, 2016. *Image source: Jessica Tovar*

cities in Los Angeles County to establish a Community Choice program with similar goals. That effort also has the strong support of several unions and has also leveraged research of the University of California at Los Angeles (UCLA).

A New Development: Community Choice in New York State

In April 2016, exercising its regulatory authority over the state's electricity system, the New York Public Service Commission (PSC) authorized the formation and operation of Community Choice Aggregation programs in New York State.[20]

In taking this action, the PSC envisioned Community Choice as a way to advance its plan to reshape the electricity system in New York State—a plan known as Reforming the Energy Vision (REV).[21] REV is a response to a number of factors. Most prominent was Hurricane Sandy and the resulting disruptions in electricity service and price escalations. REV is also an attempt to address an aging electrical infrastructure, price gouging by energy service companies (ESCOs) in the state, and the call for more renewable energy.

New York's REV is a bold effort to create a more modern, reliable, and efficient energy system for the state. It intends to augment the state's centralized fossil fuel electricity system with distributed energy resources (DERs), including renewable energy, energy efficiency, storage, and demand response technologies.

Within this vision, investor-owned utilities continue to play a dominant role. However, rather than pursuing their traditional business model of state-guaranteed return on infrastructure investments, the utilities will derive revenues from their role as distributed service platform providers, collecting fees from the DER services they support.

To make all this viable, REV is hoping to spur the development of a large new market for DER services, with a significant expansion in the number of ESCOs that provide renewable energy, energy efficiency, energy storage, automated control technologies, and other products to millions of electricity consumers across the state. In this way, REV anticipates the DER market tidal wave that is beginning to sweep across the country—what the California Distributed Energy Summit calls "a maelstrom of change that will have profound impacts on strategic directions and (business and procurement) opportunities for utilities, solar PV, energy storage, demand response, and customers on multiple fronts."[22]

Enter Community Choice. Within this utility-supporting DER market growth strategy, REV sees Community Choice as playing a key role: by aggregating electricity customers on a municipal scale, the ESCOs no longer need to market to millions of individual consumers. Instead, they can enter into contracts with Community Choice administrators who are acting on behalf of energy customers within their respective jurisdictions.

The April PSC order establishing Community Choice aggregation in New York defines the role of aggregators as brokers for services—"where the municipality acts as an aggregator and broker for the sale of energy and other services to residents but does not take ownership of the energy itself."[23] This language implies that Community Choice aggregators play a limited role: they do not represent new public energy service providers who could develop new energy resources in the community; create programs to expand community ownership and control of energy; design community-scale decentralized energy systems; or as public energy agencies address the economic, environmental, and equity needs of their communities. Rather, Community Choice aggregators are designated as middle parties, or brokers, for mass marketing energy services in a rapidly growing DER marketplace.

This limited role appears to be reinforced by financial constraints on the Community Choice aggregator. In return for brokering service contracts the Community Choice administrator can collect a fee negotiated for each such contract. However, the Community Choice aggregator cannot establish a tariff on customers' energy bills to pay for a broader program.[24] If the Community Choice aggregator is not a load-serving entity—if it does not sell electricity or services to customers—it cannot set rates or collect revenue. Under these circumstances, the Community Choice aggregator has no apparent financial base for designing and developing energy assets in the community or for creating new programs for spurring local energy resource development.

Despite the limitations of the PSC order establishing Community Choice in New York State, many Community Choice Version 2.0 advocates are working to broaden and extend the role of Community Choice energy programs through the state's Clean Energy Standard (which suggests opportunities for Community Choice "self-initiated power purchase agreements with renewable energy generators or deployment of renewable energy resources"[25]) and through the state's Clean Energy Advisory Council.

One possible approach would be for a Community Choice aggregator to become an ESCO, thereby making it a load-serving entity that can enter into power purchase agreements for local energy development or directly develop and own renewable energy assets. Perhaps the solution is an ESCO of a new type—a public, not-for-profit ESCO that has the latitude to function as a Community Choice aggregator.

In any case, Community Choice is in its infancy in New York State, and many are trying to make it a vehicle for the design and implementation of DER systems at a municipal scale.

The Significance of Community Choice in a More Unified Energy Democracy Movement

Community Choice energy programs can provide a powerful vehicle for democratizing municipal-scale energy systems. As this chapter has attempted to point out, to capture this potential requires mobilizing the community to shape the Community Choice program to provide economic, environmental, and equity benefits to the community.

Accordingly, a Community Choice program can be the basis for community engagement on the individual, group, and neighborhood levels. The program can encourage and promote community-based energy initiatives such as behind-the-meter installations, energy cooperatives, shared renewable systems, microgrids, and other collective and neighborhood-based energy projects that strengthen community ownership and control of energy. It can also develop municipal-scale projects and build publicly owned energy assets. In this way, a Community Choice program can serve as a development platform for democratizing energy and strengthening community resilience, especially in those communities hardest hit by the economic and environmental impacts of climate change.

Community Choice programs of this kind can also serve as the leading edge for transforming electricity systems at a statewide level. In California, for example, Community Choice is increasingly seen as a preferred electricity procurement model by communities across the state.

This shift draws into question the traditional role of state-regulated monopoly utilities as being the providers of last resort and guarantors of electricity system reliability. With the conditions of departing load represented by the expansion of Community Choice programs, alternative approaches to addressing system reliability are needed, and new public institutions for managing the electricity grid are being called for. The proliferation of Community Choice programs is not only democratizing energy at the municipal power level but also challenging the power and control of the monopoly utilities at the state level.

We know that with the rush to develop huge new markets for distributed energy resources, as is taking place in New York and California, that our communities can be easily bypassed—if not exploited—by this technological tsunami, unless the power of the monopoly utilities is sharply curtailed and the power of public institutions serving our communities greatly enhanced.

The democratization of municipal-scale power and the creation of a new decentralized energy paradigm is key to meeting this challenge. Not only can it unleash our communities to take control of their energy futures at the local level, but it is essential to building the public institutions, resources, leadership, and vision needed to shape and control our electricity systems at the state level.

This, in turn, advances our ability to strengthen energy democracy nationally, by demonstrating the power of community-based renewable energy development to address the economic, environmental, and equity needs of our communities.

Clean Power to the People!

Appendix A: Proposed Goals for an Alameda County, California Community Choice Program[26]

We need bold action to address escalating climate destabilization and increasing economic hardship and inequality in our communities.

An Alameda County Community Choice Energy program that prioritizes and invests in the development of local renewable energy resources can be a powerful tool to reduce greenhouse gas emissions, speed up the switch to 100% renewable sources of energy, and address equitable economic development. Investing in local clean energy development builds wealth in our communities and helps create family-sustaining jobs. County-managed development projects can increase union participation in the renewable energy sector and offer opportunities to disadvantaged job seekers in Alameda County.

We seek to establish a Community Choice program that serves the residents and businesses of Alameda County in the following ways (not in any order of priority):

1. Provides competitively priced electricity to customers, at more stable and lower rates than Pacific Gas & Electric (PG&E).

2. Prioritizes the development of local renewable resources, including reduced energy consumption and renewable electrical generation, with goals of at least 18% reduction in electricity demand through conservation and energy efficiency, and at least 50% of renewable energy being locally generated, all within ten years of the start of the program.[27]

3. Achieves Alameda County's Climate Action Plan greenhouse gas reduction goals and comparable goals of all participating jurisdictions while also exceeding the California renewable portfolio standard (RPS) and the renewable portfolio of PG&E.

4. Generates family-sustaining, high-quality clean energy jobs through local renewable resource development that prioritizes union jobs, spurs local workforce development, overcomes barriers to employment in historically disadvantaged communities, and includes local small businesses, diverse business enterprises,[28] and cooperative enterprises.

5. Promotes local and community ownership and control of renewable resources, spurring equitable economic development and increased resilience, especially in low-income communities and communities of color, which are most impacted by climate change.

6. Improves community health and safety by reducing pollution from fossil fuel power generation and by electrifying vehicle transportation.

7. Includes community stakeholders in the decision-making process of the Community Choice program and ensures inclusive representation.

12/14/14

Appendix B: East Bay Clean Power Alliance Vision: A Community–Development–Focused East Bay Community Energy Program[29]

We envision an East Bay Community Energy (EBCE) program in Alameda County that prioritizes the development of local renewable energy resources (both demand reduction and new generation) as a way to meet its stated community benefit goals.

This kind of community development–focused Community Choice program differs in some substantial ways from an investor-owned utility. In addition, this kind of Community Choice program would include a number of features to help it meet its community benefit goals, as outlined in table 8.2.

11/18/15

Table 8-2—Features of a community development–focused Community Choice energy program and the impacts these features would have on meeting community benefit goals

Feature	Impact
Implements a local build-out plan for renewable energy resource development: • Builds and integrates local renewable resources • Integrates local resources with market procurement • Specifies 10-year build-out scenarios to meet portfolio targets • Identifies financing/capitalization requirements, sources, and mechanisms, as well as return on investment	**Builds an economic development platform, which includes:** • greenhouse gas reduction • clean energy jobs • rate stability • social equity • local ownership and control of energy • community resilience • other community benefit goals
Flattened electricity load profile: Reduces/spreads out peak loads: • Uses load data to identify sources of peak loads • Designs programs to reduce/eliminate peaks • Uses storage, demand response, etc., to shift peak loads	Significantly lowers overall system costs of electricity by reducing expensive peak-load electricity
Reduced overall electricity load: • Promotes conservation, energy efficiency, demand response, building retrofits, etc. • Provides energy efficiency services for commercial, residential, nonprofit, and public buildings and monetizes the savings • Creates neighborhood-based programs to foster ratepayer consciousness of electricity consumption/waste • Promotes building retrofit financing for low-income property owners and multifamily residences	• Saves money for ratepayers • Increases economic development in energy efficiency and demand reduction • Reduces local greenhouse gas emissions • Creates local clean energy jobs • Increases social equity • Increases community energy consciousness
High renewable portfolio content: Exceeds California renewable portfolio standard (RPS)	Maximizes reduction of greenhouse gas emissions

continued on page 168

Table 8–2 continued

Feature	Impact
High local renewable portfolio content: Prioritizes community-based renewable generation • Identifies prospective sites for development and initiates development projects • Invests in building local assets • Builds technical capacity of new local businesses as renewable energy project developers and contractors • Incentivizes cooperatives, minority businesses, and collective enterprise development	• Reduces local greenhouse gas emissions • Increases local business development • Creates local clean energy jobs • Builds a more reliable, disaster-secure, and resilient electricity system • Increases social equity • Stabilizes electricity rates
Integrated power planning: Schedules local and market-purchased power to lower costs, hedge against market volatility, and provide adequate reserves	• Increases reliability • Lowers overall system costs of electricity
Promotes behind-the-meter development of energy efficiency and renewable generation resources: • Markets behind-the-meter services and financing to building owners • Creates neighborhood- or sector-based programs to promote building upgrades • Establishes easy financing mechanisms	• Increases local ownership of energy • Increases social equity • Increases community resilience • Saves money for building owners
New programs to incentivize local build-out: • Prices for excess net-metering production that encourage maximum rooftop installation • Feed-in Tariff program for new generation • Shared renewables program (virtual net metering) • PACE financing • On-bill repayment • Streamlined solar permitting for all participating municipalities • Incentives for demand response implementations	• Reduces local greenhouse gas emissions • Builds local business • Generates ratepayer savings • Increases social equity • Increases local ownership of energy assets • Creates local clean energy jobs

Feature	Impact
Labor, workforce development, and performance standards for East Bay Community Energy (EBCE) projects: • Negotiates EBCE Community Workforce Agreements (or project labor agreements) • Aggregates large numbers of small projects into larger projects done under Community Workforce Agreement • Builds pathways for local residents and disadvantaged communities into family-sustaining jobs	• Improves wages and benefits for clean energy jobs: increasing skill level of workers, increasing union jobs, and building union strength • Stabilizes communities • Increases social equity • Reduces costs of unemployment, crime, health care, and other safety net programs
Experimental/pilot programs for new technologies: • Microgrid development • New local business development • Neighborhood involvement • Partnerships with water districts Partnerships with transportation agencies	• Increases new business development and innovation • Increases local economic development • Creates local clean energy jobs • Builds a more reliable, disaster-secure, and resilient electricity system
Builds synergy with electric vehicles for public transportation, goods movement, private travel, etc.	• Lowers pollution and greenhouse gas emissions • Makes optimum use of resources • Improves local health
Creates social equity through programs to benefit communities most impacted by environmental and economic injustice: • Incentive programs and financing tailored to needs of low-income communities • Local hire and workforce development programs for disadvantaged communities • Minority and small business development programs • Opposition to utility shut-offs	• All neighborhoods benefit from energy resource development and improved environmental health • Historically disadvantaged communities benefit from local business growth and clean energy employment development
Promotes community participation in shaping and implementing the EBCE program.	• Empowers communities • Increases social equity • Increases democracy

Source: East Bay Clean Power Alliance Vision: A Community-Development-Focused East Bay Community Energy Program, http://www.localcleanenergy.org/files/EBCEProgramVision.pdf.

1. PG&E news release, June 21, 2016, https://www.pge.com/en/about/newsroom/newsdetails /index.page?title=20160621_in_step_with_californias_evolving_energy_policy_pge_labor _and_environmental_groups_announce_proposal_to_increase_energy_efficiency_renewables _and_storage_while_phasing_out_nuclear_power_over_the_next_decade.

2. Quoted in Al Weinrub, "Energy Democracy: Inside Californians' Game-Changing Plan for Community-Owned Power," *Yes! Magazine*, November 12, 2016, http://www.yesmagazine .org/new-economy/energy-democracy-inside-californians-game-changing-plan-for -community-owned-power-20151112?can_id=7cfb313905fcca33103fce281be3f17e&source =email-energy-democracy-stopping-tpp-distributed-solar-and-building-a-peoples-climate -agenda&email_referrer=energy-democracy-stopping-tpp-distributed-solar-and-building -a-peoples-climate-agenda.

3. "Meet the Latest Disruption for Utilities: Community Power," *EnergyWire*, June 9, 2016, http://www.eenews.net/stories/1060038517. An interactive map of Community Choice initiatives in California is available at the Clean Power Exchange website, http://cleanpowerexchange.org/california-community-choice.

4. Revenue bonds are repaid through revenues generated by public investment rather than through increased taxes.

5. According to the Santa Rosa Climate Action Plan, for example, the authors determined it would be extremely difficult for the city to meet its climate action goals unless it dealt directly with electricity consumption, and this was one of the rationales for the city council's voting to join Sonoma Clean Power.

6. A net metering program charges customers who have behind-the-meter (rooftop) solar facilities for net energy they consume from the grid and credits them for any net energy they generate into the grid.

7. A feed-in tariff program incentivizes new renewable energy generation through standardized purchase contracts that guarantee a set payment for all generated electricity for a set duration of time (usually twenty years)

8. A shared renewables (usually solar) program allows for multiple investors and/or subscribers of a renewable energy–generating facility to share the benefits of the electricity generated as a way for renters and others unable to own their own solar system to reap the benefits of a solar generating facility.

9. Two days after the historic 2014 People's Climate March in New York City calling for climate action, federal and California State officials released an 8,000-page proposal for private renewable energy development on 22.5 million acres of California desert. See Carolyn Lochhead, "Energy Plan Calls for Big Renewables Projects in State's Deserts," September 23, 2014, http://www.sfgate.com/green/article/Sprawling-solar-farms-OKd-near-desert -national-5775871.php.

10. There are many studies that reflect the technical potential of decentralized energy systems. For example, see *U.S. Renewable Energy Technical Potentials*, National Renewable Energy Laboratory, July 2012, http://www.nrel.gov/docs/fy12osti/51946.pdf, and *Bay Area Smart Energy 2020*, March 2012, http://pacificenvironment.org/-1-87.

11. For detailed arguments about the benefits of decentralized energy systems, see *Community Power: Decentralized Renewable Energy in California*, http://communitypowerbook.com.

12. For an explanation of renewable energy certificates and their relationship to Community Choice, see *What the Heck Is a REC?*, http://www.localcleanenergy.org/what-the-heck-is-a-rec.

13. Paul Fenn, a founder of the Community Choice movement and author of California's

Community Choice law, was instrumental in drawing the distinction between Community Choice Version 1.0 and Version 2.0.

14. "Behind the meter" refers to the customer's side of an electricity meter (as opposed to the grid side): electricity generated or demand reduced on-site, so it is not measured by the meter (for example, rooftop solar generation, energy efficiency upgrades, Energy Star appliances, and so forth).

15. Program by which loans to homeowners or business owners for solar installations or energy efficiency retrofits are paid back over time through their property tax bills.

16. Incentive program that allows customers to pay off the initial cost of a home solar installation or energy efficiency retrofit through their monthly utility bill.

17. *California's Distributed Energy Future, Fireside Chat*, March 16, 2016, timestamp 12:37–13:38, http://www.greentechmedia.com/multimedia/view/fireside-chat: "One of the bigger shifts that we see at the policy level is, is people clamoring for these clean community aggregators. . . . These CCAs are really just a coup. It's local governments making decisions to carve off a piece of the customer [base] and sort of in a forced collectivization."

18. Ibid,, timestamp: 10:32–12:27.

19. California Alliance for Community Energy website, http://cacommunityenergy.org.

20. NYS Public Service Commission, *Order Authorizing Framework for Community Choice Aggregation Opt-Out Program*, April 21, 2016, http://documents.dps.ny.gov/public/Common/ViewDoc.aspx?DocRefId=%7B38EFD3B0-48BC-400E-9795-98CB5EFAE0FA%7D.

21. NYS Department of Public Service Staff Report and Proposal, *Reforming the Energy Vision*, April 24, 2014, http://www3.dps.ny.gov/W/PSCWeb.nsf/96f0fec0b45a3c6485257688006a701a/26be8a93967e604785257cc40066b91a/$FILE/ATTK0J3L.pdf/Reforming%20The%20Energy%20Vision%20(REV)%20REPORT%204.25.%2014.pdf.

22. California Distributed Energy, "Seize Opportunities in California's Emerging DER Market," September 2016, http://infocastinc.com/event/california-distributed-energy.

23. NYS Public Service Commission, *Order Authorizing Framework*, p. 49.

24. Ibid., p. 36.

25. Ibid., p. 37.

26. East Bay Clean Power Alliance, December 2014, http://www.localcleanenergy.org/files/EBCPA_DraftGoals%2Bpreamble_12-14-14-2.pdf.

27. Targets taken from scenario in Al Weinrub and Seth Baruch, *East Bay Community Choice Energy: From Concept to Implementation*, February 2014, http://www.localcleanenergy.org/files/Community%20Choice%20Energy%20in%20East%20Bay.pdf.

28. Includes minority-owned, women-owned, and disabled veteran–owned businesses, and other such enterprises

29. East Bay Clean Power Alliance, November 2015, http://www.localcleanenergy.org/files/EBCEProgramVision.pdf

CHAPTER 9

Community-Anchor Strategies for Energy Democracy

MAGGIE TISHMAN

Anchor institutions—large, place-based entities like hospitals, universities, cultural attractions, public housing, and other major public institutions— play an outsized role in their local economies. As large consumers of energy, drivers of infrastructure investments, and major political players, they are particularly influential in the energy sector and will play a critical role as this sector shifts to more renewable and distributed generation.

Communities seeking a more democratic and sustainable local economy can benefit from enlisting anchor institutions as partners in a just transition. Overcoming power asymmetries and influencing long-term investment decisions require creativity, persistence, and patience but can result in considerable community benefits and opportunities for self-determination.

This chapter tells the story of two community groups in the Bronx that are successfully developing meaningful and impactful community-anchor partnerships in the energy sector. These groups are part of a larger effort to democratize the Bronx economy—known as the Bronx Cooperative Development Initiative—and have chosen the energy sector because of the opportunities it presents for reimagining how decisions are made and who benefits.

Mothers on the Move (MOM) is leveraging the New York City Housing Authority's energy performance contracts to create economic opportunity for public housing residents, and the Northwest Bronx Community and Clergy Coalition is partnering with Montefiore Medical Center to link energy efficiency upgrades for apartment buildings with building-wide asthma interventions. With these stories, we hope to inspire and guide other communities to work with their local anchor institutions to model energy democracy.

We begin our story with a snapshot of the Bronx, before examining the role of anchors in the local economy and the emerging opportunities for community-anchor partnerships in the energy sector. We then highlight the two specific projects mentioned above, in which Bronx community groups are leveraging anchors' investments to achieve physical resilience, economic development, and community health for residents. Finally, we consider some of the challenges of forging community-anchor partnerships, especially in the energy sector, and identify a potential path forward to build and maintain these transformative relationships.

Background

The Bronx, like many urban areas, is home to tremendous assets, including: cultural, as the birthplace of hip-hop; natural, with the most trees of the five New York City boroughs and the only source of freshwater; and built, with an abundance of large, flat rooftops that could be used for solar panels.

Also, like in other urban communities, many resources reside in so-called anchor institutions—large, place-based institutions with significant fixed assets in a community.[1] In the Bronx, these include Montefiore Medical Center, among the largest hospitals in the Greater New York region; the New York City Housing Authority, which manages over 44,000 units in the Bronx; and, among others, the Hunts Point Food Distribution Center, the New York Botanical Gardens, Fordham University, the Bronx Zoo, and St. Barnabas Hospital.

Collectively, residents hold many resources as well: their purchasing power, their political might, their dense social networks, and their strong community-based organizations. However, decades of disinvestment, concerted defunding of local planning capacity, and structural racism have left community members with many social and economic challenges and without access to the resources within anchor institutions that could help generate local wealth and improve quality of life.

The energy sector encapsulates this dynamic clearly. Because of the essential role it plays in modern life, Bronx residents must consume energy. Yet energy is a considerable economic burden for low-income Bronx households; almost all money related to this important economic sector currently leaves the borough; many residents' energy supply is vulnerable in the event of a major storm; and traditional energy generation methods produce greenhouse gases and other harmful pollutants that contribute to climate change and respiratory illness.

Many Bronx anchor institutions are now recognizing the importance of sustainability and resiliency to their lines of business and are planning major energy investments that will affect residents, but often with little community consultation or coordination. Thus, an influx of investment in renewable, clean, and resilient energy may do little to address residents' economic and environmental challenges.

New technological developments and shifting government regulations are opening up new opportunities to localize control of energy production, spur adoption of clean and renewable energy, and increase residents' decision making in public infrastructure investments. However, harnessing the demand for energy and related services from anchors to drive local economic development and other community benefits requires intentional and concerted action from community groups and balanced partnerships between communities and anchors.

Mothers on the Move and the Northwest Bronx Community and Clergy Coalition are two member organizations within the Bronx Cooperative Development Initiative (BCDI)—a network first convened and launched by the Massachusetts Institute of Technology Community Innovators Lab (MIT CoLab). BCDI seeks to create *economic democracy* by organizing the collective power of residents, building deep and meaningful relationships among diverse stakeholders, and implementing transformative projects that generate wealth and promote self-determination for local residents. Individually and collectively, BCDI members are reorienting the relationships between community and anchor institutions and leveraging anchors' investments and influence to win significant community benefits.

Central to the work of the Bronx Cooperative Development Initiative is learning about alternative, cooperative economic models. In 2015, members of BCDI traveled to Mondragón, Spain—home to the largest network of worker-owned cooperatives in the world—to learn about economic democracy (figure 9-1). Representatives from two BDCI initiatives—Mothers on the Move and the Northwest Bronx Community and Clergy Coalition—were among the delegation, many of whom have since translated Mondragón's principles into their work in the energy sector and beyond.

The Power of Community-Anchor Partnerships

Hospitals, universities, cultural attractions, and major public institutions are all examples of anchor institutions. These entities have significant fixed assets in a given location, are unlikely to move, and so will play a role in their local economies for the foreseeable future.

Figure 9-1. Members of the Bronx Cooperative Development Initiative traveled to Mondragón, Spain, in 2015 to learn about cooperative development models. *Image source: BCDI*

Historically in the United States, many anchor institutions were built in the urban core, and as domestic manufacturing declined, anchor institutions became some of the largest employers in their respective regions.[2]

As people and capital moved to the suburbs in the second half of the twentieth century, the so-called "eds and meds" or MUSH (municipalities, universities, schools, and hospitals) market came to control vast amounts of capital in disinvested urban neighborhoods. This dynamic has naturally created tension between often-impoverished local residents and resource-rich anchors; however, over the past two decades, anchors have increasingly understood their ability to lift up the surrounding community, and residents have increasingly approached anchors as valuable neighbors and partners.[3]

ANCHORS AND THEIR ROLE IN THE LOCAL ECONOMY

Following are just some of the ways in which anchors can support the local economy and provide targeted community benefits:

- **Employment.** Anchors are among a region's largest employers. Partnering with local public schools, providing training, and creating on-ramps for existing residents are all ways that anchors can support local employment.
- **Purchasing.** Beyond employment, anchors spend large amounts of money on goods and services. However, they often rely on regional, national, or international corporations and bypass local businesses to secure these products. Redirecting this spending locally can make a significant impact on the economy, building both income for workers and wealth for local business owners.
- **Direct community investing.** Anchors can also invest directly in community projects and programs. Because they are rooted in place, anchors can themselves benefit from improved health and safety in their neighborhoods and simultaneously bolster their own image among community members. Investment can take many forms, including affordable housing, green space, and schools.

- **Sustainability.** Large anchor institutions play a particularly important role when it comes to sustainability. They are large consumers of energy, and if they do not effectively manage their carbon footprints, anchors can be major polluters, producing emissions that affect community health. They also have the physical assets, financial capital, and political influence needed to host large-scale clean and renewable energy resources that can service the entire community.
- **Resilience.** Last, anchors often serve as critical infrastructure, providing refuge and community services after a disaster. They have the resources (and incentive) to invest in systems that protect themselves and surrounding communities from future disasters. Beyond physical resilience, hiring and purchasing locally and investing in the community can support economic and social resilience, which are equally crucial to a community's ability to weather setbacks.

EMERGING OPPORTUNITIES IN THE ENERGY SECTOR AND BEYOND

Because of the size of most anchor institutions, energy expenses can factor large in their budgets. Increasingly, business and nonprofit managers are paying more attention to the role of energy in their bottom line and are managing it strategically to lower costs and maximize efficiency. Many anchors are also concerned about resiliency, particularly in New York City, where Hurricane Sandy caused major physical damage in 2012. The storm cut power at 400 New York City Housing Authority buildings, and heat and hot water at 386 buildings, stranding some 80,000 tenants without these basic services. The damage to that agency's portfolio alone totaled over $3 billion. Montefiore Medical Center was not affected, but other hospitals had to rely on backup generators and, in some cases, evacuate patients when that emergency power failed.

In New York State, a changing regulatory landscape is facilitating and incentivizing local energy generation, which can be cleaner, more efficient, and more resilient. Many anchors are considering efficient co- and trigeneration technologies (which produce heat, electricity, and sometimes cooling simultaneously) or fully fledged microgrids, which can operate independently in the event of a major power outage. In New

York City, in particular, Mayor Bill de Blasio committed to reducing the city's carbon emissions by 80% compared with 2005 levels by 2050 and has developed corresponding plans to retrofit public anchor institutions, including public housing, hospitals, and schools.

Public anchors are particularly important in achieving energy democracy because city agencies often have goals for hiring and sub-contracting with women, people of color, and other disadvantaged residents. Holding public agencies and general contractors accountable to meeting these goals can ensure that investments in sustainability and resiliency result in economic development for historically disadvantaged communities.

Beyond government institutions, regulatory changes in the health-care sector are changing investment patterns for nonprofit hospitals, making these anchors particularly good partners in the movement for energy democracy. Thanks to provisions in the 2010 Affordable Care Act, nonprofit hospitals must invest in their local communities in order to maintain tax-exempt status.

Medicaid reform has also incentivized these anchors to address the upstream causes of poor health, collectively known as the social determinants of health. Some of these determinants, like poverty and inadequate housing conditions, can be addressed through energy efficiency retrofit or green jobs programs. Medicaid funds can support these initiaves, rather than simply reimbursing hospitals for poor patients' emergency care. In states like New York, where Medicaid reform is already underway, even repealing the Affordable Care Act is unlikely to change this long-term shift in investment patterns.

Many community groups[4] have realized, and are taking advantage of, the opportunities for greater self-determination and economic development within the energy sector (and health sector). A growing *energy democracy* movement—in New York state and beyond—is calling for and prototyping a system that recognizes community residents not just as energy consumers but as innovators, planners, decision makers, and implementers to achieve an energy system that is equitable, sustainable, and democratic.

Two Communities Lead the Way

Two examples show the power of community-anchor relationships in the Bronx.

ENERGY PERFORMANCE CONTRACTING FOR ECONOMIC DEVELOPMENT: MOTHERS ON THE MOVE AND THE NEW YORK CITY HOUSING AUTHORITY

Home to over 600,000 people, the New York City Housing Authority (NYCHA) is the largest public housing agency in the country. Comprised of dense campuses of high-rise buildings that span multiple city blocks, NYCHA is, in many ways, a city unto itself. And in neighborhoods with concentrated public housing, like the South Bronx, NYCHA is a major driver of community health, jobs, and real estate development.

NYCHA was founded in the 1930s and subsequently expanded to house the families of veterans returning from World War II. However, following a national pattern of disinvestment in services for the poor, NYCHA buildings began to seriously deteriorate by the 1970s. Today, aging buildings, coupled with chronic underfunding, have left NYCHA unable to effectively manage its 2,553-building portfolio. Buildings are plagued with mold, lead, pests, broken elevators, leaks, and structural deficiencies. Yet, they remain a critical resource for the low-income New Yorkers who are able to secure one of their subsidized apartments, and the community centers within many of the developments are critical infrastructure that serves as refuge in the event of a disaster.

Within that context, the community-based organization Mothers on the Move (MOM) has been organizing NYCHA tenants to win improved services and building upgrades for over ten years. Mothers in the Hunts Point and Longwood neighborhoods started the organization in 1992 to win educational reforms, but MOM later expanded to tackle housing and environmental justice issues across the South Bronx as well. With allied groups, MOM has previously fought for and won the closure of a local fertilizer plant whose toxic emissions were sickening nearby residents, and $97 million in state funding to convert the elevated Sheridan Expressway into a surface boulevard.

MOM has also helped tenants sue NYCHA to win needed repairs to their buildings. In 2009, when the American Recovery and Reinvestment Act allocated substantial funding for energy efficiency and green infrastructure, MOM interviewed residents and published a white paper that called on city and state officials to retrofit public housing and launch a South Bronx Institute for Green Careers that would provide paid technical education and job placement services.[5] MOM also talked to NYCHA residents about small businesses they could form to support sustainable practices, such as recycling.

NYCHA did not immediately take up these suggestions; however, the agency has since become interested in energy efficiency as a way to manage expenses. NYCHA provides heat and hot water for tenants, and for the vast majority of its developments, NYCHA also pays for tenants' electrical usage. During summer months, when many residents use electric air conditioners to cool their apartments, usage can run high. Effectively managing energy usage can greatly reduce NYCHA's operating expenses and capital expenses by extending the life of already installed equipment. In addition, cold or humid weather and heating-related emissions can trigger asthma and other chronic health conditions, so better climate control can increase residents' comfort and health.

In May 2015, the agency released the NextGeneration NYCHA (NextGen NYCHA) report, which outlines changes to infrastructure investing, property management, and resident engagement. Building on Mayor de Blasio's commitment to reduce carbon emissions by 80% over 2005 levels by 2050, NYCHA began moving forward with a program of energy performance contracting to reduce energy costs and carbon emissions. This meant that NYCHA would contract with an energy service company (ESCO) to provide energy efficiency improvements, to be paid back by the realized energy savings.

MOM saw an opening to reintroduce their collaboration proposal. The energy performance contract helps NYCHA reduce energy costs in two ways: (1) by installing more efficient equipment, including lighting and heating systems, and (2) by providing training to residents to manage and reduce their energy usage. MOM knew that they were uniquely equipped to train residents, and through their partnership with a local energy equipment manufacturing company, Intech 21, Inc., they could even locally assemble and sell efficient equipment.

Moreover, Section 3 regulations from the U.S. Department of Housing and Urban Development (HUD) require NYCHA to hire local workers and contract with disadvantaged business enterprises. NYCHA also signed a project labor agreement that sets goals for hiring and subcontracting minorities and women. If MOM could help connect NYCHA tenants and people of color to jobs—either manufacturing and installing equipment or training other residents—MOM could help NYCHA meet these goals.

The organization knew it had to act quickly. Including MOM as a subcontractor would help energy service companies distinguish themselves as they bid on NYCHA work. Thus getting commitments from these companies early on, before NYCHA awarded the contract, was key. MOM ultimately chose to partner with two companies—Ameresco and Constellation—each of which won energy performance contracts with NYCHA for a different set of housing developments.

MOM forged this nontraditional partnership (figure 9-2) to leverage NYCHA contracts with energy firms to ensure that public housing residents are included in the clean energy future as smart consumers, producers, and, ultimately, advocates. Specifically, to reduce its energy usage, NYCHA is contracting with two energy service companies (ESCOs), which are, in turn, subcontracting to MOM to train tenants on energy-saving techniques. MOM will also help tenants understand their role in the energy system and how they can mobilize to take more control. MOM will hire tenants to conduct the trainings and, with the energy technology company Intech 21, will explore creating a tenant-owned equipment assembly business. The tenant education funds allocated for MOM build residents' clean energy knowledge and workforce and business skills, as well as the community power needed to persuade NYCHA to continue investing in clean and efficient infrastructure and, potentially, resident ownership going forward.

Looking to the longer term, MOM and Intech21 have also taken steps to create a local business that assembles, sells, and installs in-unit tem-

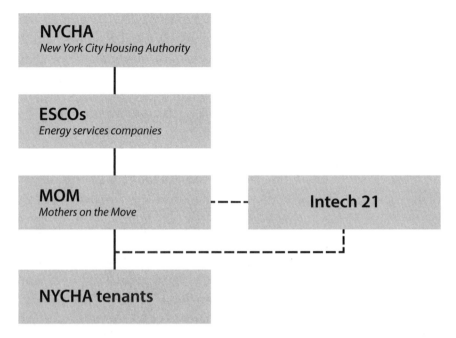

Figure 9-2. MOM's nontraditional public, private, and community partnership increases participation of public housing residents in the clean energy economy. *Image source: BCDI*

perature sensors and wireless energy modules that can monitor temperature data over time and provide ongoing training and recommendations to further increase efficiency. Most important, the business will be organized as a shared-wealth enterprise, to create jobs and wealth for low-income Bronx residents, including NYCHA tenants. The business can begin with NYCHA as a client, using its relationship with NYCHA to advance such investments, and can expand to other public housing agencies or private multifamily buildings within and beyond New York City.

MOM's work illustrates how communities can leverage public anchors' investment in energy efficiency to achieve community economic development. Their strategy takes advantage of federal and local requirements to hire local and minority workers and to contract with disadvantaged business enterprises, and it positions the deep relationships that community groups have with local residents of color as an asset to achieve those goals. Importantly, it creates a virtuous cycle whereby

NYCHA developments will become more sustainable (financially and environmentally), residents will enjoy more comfortable homes, and the economic benefit from NYCHA's investment can stay with the residents the agency serves.

THE BRONX HEALTHY BUILDINGS PROGRAM: THE NORTHWEST BRONX COMMUNITY AND CLERGY COALITION AND MONTEFIORE MEDICAL CENTER

As the Mothers on the Move—New York City Housing Authority example shows, communities can leverage energy investments to achieve greater economic development. With some additional creativity, they can also use energy investments to secure other community benefits, such as better health.

Flanked by the New York Botanical Gardens and the Harlem River and bisected by the aptly named Grand Concourse, the Northwest Bronx is, in some ways, a world apart from the South Bronx. While the South Bronx includes a number of industrial uses and major trucking routes, the Northwest Bronx is a mostly residential area with several retail corridors interspersed.

When disinvestment and arson famously engulfed the South Bronx in the 1970s, residents and faith leaders in the Northwest Bronx organized to form the Northwest Bronx Community and Clergy Coalition (NWBCCC). With the motto "Don't move, improve!" NWBCCC began organizing tenants to improve their housing conditions. As an extension of this work, in the 1980s, NWBCCC began administering federal grants for the Weatherization Assistance Program to help make low-income homes more energy efficient and comfortable.

As the organization grew, NWBCCC also adopted a number of campaigns and issues beyond housing, including hospital closures, schools, and economic development. One of their biggest wins was a twenty-year fight to redevelop the Kingsbridge Armory—a humongous, castle-like structure in the heart of the Kingsbridge neighborhood. The community benefits agreement they eventually signed with the developer was a landmark victory and has become a model for communities across the country seeking to capture investment for local economic development.

As the community benefits agreement demonstrates, NWBCCC's community organizing had become very sophisticated and powerful over the decades. However, over that same period, their members had grown increasingly poor relative to the rest of New York City. Organizing to save a hospital from closing or even to get a community benefits agreement for new development is, by its very nature, reactive and does little to reorder the economic and political power imbalances that make such fights necessary in the first place.

Meanwhile, staff and leaders at NWBCCC felt that their successful weatherization program, which served tens of thousands of residents in their catchment area, did not connect to their tenant organizing and leadership development activities. This was a missed opportunity to build a movement around energy systems and energy democracy.

In 2011, the Northwest Bronx Community and Clergy Coalition participated in a Bronx-wide development study supported by the Massachusetts Institute of Technology Community Innovators Lab (MIT CoLab),[6] which eventually led to the launch of the Bronx Cooperative Development Initiative. Through that process, community-based organizations across the borough identified energy and health as two key sectors with opportunities for economic development and shared wealth creation. NWBCCC began to see how they could both shift the balance of power within these systems and expand their building-wide retrofit work to further develop residents' leadership.

As a result of the study, MIT CoLab trained a group of forty NWBCCC leaders on economic democracy. This taught them how the local economy works and how they could intervene to ensure they were decision makers and beneficiaries of future economic development. Suddenly, the organization's theory of change became clearer: they must not only prepare leaders to fight bad development but should also prepare leaders to recognize existing local assets and develop the economy in the way they envision—especially in the energy and health sectors.

NWBCCC thus began organizing faith institutions to retrofit their buildings as a strategy to save money and protect the environment. By linking faith institutions to Bronx-based contracting companies, they could support local green jobs and businesses as well. This program increased the literacy of staff and leaders in energy democracy, both in theory and in practice.

Around the same time, NWBCCC also began learning about provisions in the Affordable Care Act that mandated nonprofit hospitals to invest in their communities as well as state-level reforms to invest Medicaid dollars in the upstream causes of poor health, such as housing. These could be major sources of investment in the community, which residents could proactively shape. The organization founded a Health Justice Committee to focus more on these issues.

Pulling these two streams of work together, NWBCCC began to plan for the Bronx Healthy Buildings Program. For decades, asthma has been a major health concern in the Bronx. Each year, there are roughly 225 emergency department visits per 10,000 Bronx residents—nearly double the rate of New York City as a whole and triple that of New York State. While the causes are complex, health professionals have long known that pests, mold, and other aeroallergens in the home can trigger dangerous and sometimes life-threatening attacks.

Because most Bronx residents are renters, they cannot directly address many of the exacerbating building conditions themselves. However, NWBCCC had experience with exactly the sorts of building-wide interventions that could help. They knew that offering landlords savings through energy efficiency could convince them to agree to health-related building interventions as well. Moreover, if construction work went to local contractors and workers, the program could grow the local green economy.

Thus, a well-designed program that targeted "hotspot" asthma buildings could address several issues NWBCCC cared about: mitigating the upstream causes of poor health, creating jobs and supporting businesses in the high-road green construction sector, and positioning residents as drivers of decisions about their home environment and their health.

An opportunity to fund such a program emerged via the BUILD Health Challenge, a consortium of funders supporting community health interventions in low-income, urban communities. As part of its grant application to BUILD Health Challenge, NWBCCC was required to find a one-to-one match from a local health-care system. This gave NWBCCC the entrée it needed to engage Montefiore Medical Center in this effort.

Montefiore, meanwhile, had their own goals to reduce asthma admissions. As part of New York State's Medicaid reform process, Montefiore and nearby St. Barnabas Hospital had partnered to form a group called Bronx Partners for Healthy Communities (BPHC), which had already determined that they would use a portion of their Medicaid reform dollars to support community health workers and integrated pest management for asthma patients. When NWBCCC approached Montefiore, they established a referral process so the Northwest Bronx Community and Clergy Coalition could access their Bronx Partners for Healthy Communities services in support of the Bronx Healthy Buildings Program. If the program demonstrated success, Healthy Buildings could become a vehicle to invest both Medicaid and community benefit dollars in the future.

Montefiore further committed to share data extracted from their electronic medical records with MIT CoLab so they could help NWBCCC identify where clusters of asthma patients existed. Combining these data with publicly available information about the size, condition, and energy usage intensity of multifamily buildings, NWBCCC could target the most problematic buildings for intervention. Comparing medical record data before and after the intervention could also be used as one measure to evaluate the program.

Last, BUILD Health Challenge required the local department of health to support the program, and NWBCCC found that the New York City Department of Health and Mental Hygiene was an enthusiastic partner, offering to lend elements from its existing Healthy Homes program. A complex hospital, public sector, and community partnership (figure 9-3) was forged to create the Bronx Healthy Buildings Program, a holistic building retrofit program.

The basic commitments were in place by June 2015, when NWBCCC was awarded the BUILD Health Challenge grant. After overcoming several challenges—including securing data-use permissions, establishing

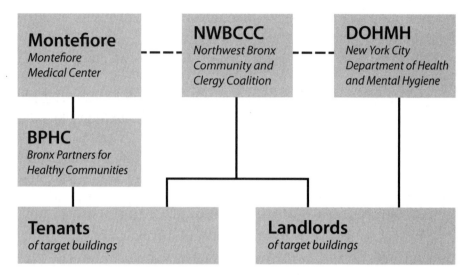

Figure 9-3. NWBCCC's anchor, public, and community partnership structure for the Bronx Healthy Buildings Program *Image source: BCDI*

a patient referral system, determining which landlords and buildings to pursue, and designing a holistic evaluation tool—NWBCCC is now moving forward with a pilot round of three buildings. These buildings will receive energy efficiency upgrades, and NWBCCC will train residents on their rights, the social determinants of health, and the connections between health and climate change. NWBCCC can also refer Medicaid asthma patients to additional resources from the health system, and with support from the Department of Health and Mental Hygiene, NWBCCC will help landlords shift to healthier building management practices, like integrated pest management and green cleaning.

In this way, Healthy Buildings uses energy efficiency as an incentive for landlords to upgrade their buildings and grant access to community organizers. It then leverages the investments of nonprofit hospital anchors and the local health department to provide health services and operational improvements. The end result is a program that lowers energy consumption, improves community health, and creates green jobs. This is one example of how communities can use both energy and anchor investments to their advantage.

Lessons for the Future

The rich experiences of BCDI in working with anchor institutions in the Bronx provide a number of lessons for carrying this work forward.

REMAINING CHALLENGES

Despite the budding potential of community-anchor partnerships, challenges persist and the experiences of MOM and the Northwest Bronx Community and Clergy Coalition highlight many of these.

- **Planning for technical fields in a truly inclusive and participatory way.** Residents need to be involved in both planning and implementation of energy projects in order to truly shift economic and political power. However, detailed project planning with a large group of leaders is often difficult to manage. NWBCCC has experienced this tension with the Bronx Healthy Buildings Program as the number of partners and complexity of the program have grown over time. Ensuring their leaders feel ownership of the project has sometimes proved difficult to balance with the need to coordinate with anchors and bring on additional outside capacity for finance and other technical aspects of the program. As a result, most leaders within the organization have not yet come to see hospitals as partners in energy democracy or come to accept community-anchor partnerships.
- **Navigating internal anchor bureaucracy.** The internal bureaucracy of anchor institutions has significantly slowed down both MOM's and NWBCCC's project. While ethical reviews are extremely important for protecting the public from potentially dangerous research, and public agencies must ensure that they are investing in the most effective and transparent ways, communities must be prepared for an arduous and, at times, frustrating process.
- **Reconciling incompatible timelines.** The longtime horizon for major anchor investments necessitates that stakeholders get involved early and stay involved through the planning and implementation process. This is a resource-intensive endeavor that requires a long-term vision and, importantly, consistent funding for leadership development.

Some funders have understood this and have put significant resources into the planning and organizing necessary to lay the groundwork for energy democracy. However, many have unrealistic expectations about measurable outcomes within a short time frame.

- **Dealing with mediating organizations**. Often, additional organizations mediate between community members and anchor institutions. These mediating organizations may respond to community or anchor needs but may also have their own agendas and priorities, inserting additional layers of complexity. For NWBCCC, Montefiore's partnership is mediated through Bronx Partners for Healthy Communities and the overall Medicaid reform process, which affects their ability to make direct investments and thus form a meaningful and direct anchor-community partnership.

  In MOM's case, trade unions and energy services companies—both of which have their own, sometimes conflicting interests—mediate between MOM and NYCHA. NYCHA must hold mediators accountable to local hiring goals they share with MOM but often must do so with insufficient information or institutional will. Even when anchors are completely aligned, they may not have complete influence over mediating organizations.

- **Overcoming distrust**. Working with anchors almost always requires overcoming distrust among leaders, based on negative past experiences. Traditional community organizing often leads to generalizations about anchor institutions or other centers of power, which makes mobilizing community members easier but building constructive solutions together with anchors more difficult. Getting community members to see anchors as allies, or at least as complex and multifaceted institutions, requires new organizing strategies and intentional relationship building.

- **Maintaining partner relationships**. In addition to Montefiore, the Bronx Healthy Buildings Program has over a dozen partners that are contributing to or advising the program in some way. Keeping partners updated in real time in order to take advantage of their expertise is extremely resource intensive, yet few resources exist to support this aspect of program management. Similarly, MOM must frequently liaise between ESCOs, NYCHA, and residents, yet there are currently no resources designated to support this role.

- **Maintaining accountability.** The power differential between community groups and anchor institutions can make it difficult for communities to hold anchors accountable to commitments. BUILD Health Challenge attempted to reconcile this imbalance by placing community groups in the lead, which helped bring anchor institutions to the table but has not necessarily held them to institutional commitments.

This presents a major tension in building effective community-anchor partnerships: Compared with a cash donation, in-kind anchor support may imply deeper institutional buy-in and longer-term partnership but also compromises full community control of the project. Meanwhile, MOM and allied groups must continue to press NYCHA to monitor and enforce inclusive hiring and subcontracting goals in order to realize community benefits.

A PATH FORWARD

Despite these challenges, community-anchor partnerships are successfully building literacy in energy democracy among community residents, shifting the power dynamic between community and anchors, and creating economic opportunities and other benefits. And these models can be replicated as the energy sector continues to shift in the Bronx and beyond. The experiences of these two community groups provide several vital lessons for other groups looking to build community-anchor partnerships to further energy democracy:

- **Use an asset-based framework.** Working with anchors is about shifting the fundamental balance of power so that community becomes an asset rather than a nuisance or liability. Asset mapping is a key strategy to achieve that shift. NWBCCC knew that the stock of old buildings in their catchment area and their experience organizing and retrofitting buildings were assets that could help Montefiore reach its goal of addressing asthma. For MOM, their ability to connect with residents and their relationship with Intech 21 positioned them as a valuable partner that could help NYCHA meet its goals. Using an asset-based framework, building community-anchor partnerships does not rely on anchors' goodwill but rather the recognition of mutually aligned interests.

- **Recognize favorable policy shifts.** Policy shifts that enable or require anchors to change their orientation to the community were key in both cases. HUD Section 3 requirements and Mayor de Blasio's commitment to greenhouse gas reduction were crucial in the NYCHA case. For Healthy Buildings, the Affordable Care Act and Medicaid reform strengthened anchors' incentives to work with community groups to improve health outcomes. Communities must recognize, advocate for, and take advantage of these favorable policy shifts. Here again, community groups can position themselves as assets, helping anchor institutions adapt to achieve goals set in place by new policies.

- **Find internal champions.** Anchor institutions are not monolithic, and finding internal champions is key. Building a personal relationship with someone in a position of power can help reorient internal decision making and resources. Equally important, community groups should seek to find and cultivate *multiple* internal champions, so that partnerships can last even as individual staff members may leave or move within the organization. NWBCCC has found two internal champions at Montefiore. For MOM, that task has proved more difficult, which, in turn, has made it harder to advance their project.

- **Get commitments in writing.** While anchors may easily express verbal support for a project, community groups should aim to secure formal letters of commitment specifying resources that anchors will contribute to the project. This can help the communities hold anchors accountable to supporting the project with resources in the future, especially if internal champions leave the organization or move to new roles.

- **Build direct relationships.** As much as possible, community groups should try to build direct relationships with anchors, with anchors directly investing their resources, rather than marshaling resources from other indirect sources. This helps avoid conflicting interests or influence from mediating organizations.

- **Invest in leadership development**. Community groups must invest in leadership development and build the necessary infrastructure to take advantage of these opportunities. The Bronx is fortunate to have a set of very savvy and sophisticated community-based organizations, and economic democracy training gave them additional vocabulary and frameworks with which to analyze emergent opportunities and propose strategies.

  For funders looking to support community-anchor partnerships, leadership development within community groups is an important first place to invest in building a balanced and meaningful partnership. And sustained investment can bring in more community leaders and deepen their engagement as the partnership launches and grows.

- **Invest in partnerships that put community groups in the lead.** These groups know their needs best and, based on the lived experience of their members, have unique insights and relationships that can help solve their problems. Investing in community groups also helps level the otherwise extreme power imbalance between anchors and community. Requiring an anchor match for grant funding can also help hold anchors accountable to community partnerships.

- **Reserve a portion of funding to support leadership development, participatory planning, and partner coordination.** As community groups partner more readily with anchor institutions, without concerted resident engagement, they risk losing connection to their base. These groups need resources to develop their leaders in technical fields like energy and health, and the resources to involve their leaders in designing partnerships and programs with anchors. Community groups may also need to bring in other technical assistance partners to supplement their local knowledge. Building and maintaining these relationships while implementing programs requires significant investment.

MOM, NWBCCC, other members of the Bronx Cooperative Development Initiative, and other groups within and beyond the Bronx will continue to engage anchors in creative ways. The rapidly changing energy sector provides a particular opportunity to do this and realize community benefits. With these stories, we hope to inspire and guide other communities to work with their local anchor institutions to model energy democracy.

1. Community-Wealth.org, "Overview: Anchor Institutions," accessed January 13, 2017, http://community-wealth.org/strategies/panel/anchors/index.html.
2. Ibid.
3. Ibid.
4. In this chapter, the terms *community* and *community groups* refer to community-based organizations that educate, mobilize, advocate for, and represent residents. The term *residents*, on the other hand, refers to individual community members who may or may not form part of these organizations but do often benefit from the changes they bring about.
5. Mothers on the Move and the Urban Justice Center, *South Bronx Residents Solution on Greening Our 'Hood*, https://cdp.urbanjustice.org/sites/default/files/cdp_10feb11.pdf.
6. Massachusetts Institute of Technology Community Innovators Lab, *Development Study for the Bronx Cooperative Development Initiative: A Model for Economic Transformation*, http://colab.mit.edu/sites/default/files/MITCoLab_BCDI_Development_Study.pdf.

CHAPTER 10

New Economy Energy Cooperatives Bring Power to the People

LYNN BENANDER, DIEGO ANGARITA HOROWITZ, AND ISAAC BAKER

Energy democracy is a cornerstone in the struggle for climate resilience and an essential step toward building a more just, equitable, and sustainable future. The growing movement to democratize energy is creating an alternative to our fossil fuel economy that puts racial, social, and economic justice at the forefront of the transition to a 100% renewable energy future. It represents the energy component of the nationwide movement for a new economy.

Cooperatives are an important part of this new economy movement. They have a deep history in the United States and around the world, and they are already bringing significant value to people and communities. Because cooperatives are an effective way communities can democratically own and control resources, new energy cooperatives are emerging across the United States in response to the need for energy alternatives.

Richard Heinberg, a world-celebrated journalist and educator writing about climate change and oil depletion, identifies the important role energy cooperatives can play in addressing our climate crisis: "Energy cooperatives put ownership of energy infrastructure in the hands of the people actually using the energy produced. Unlike publicly traded corporations, cooperatives don't have to pursue increased shareholder value above all; they can strive for other goals, like sustainability and equity. Energy cooperatives are vital for the transition to renewable energy—which, to be successful, will have to move faster than the market alone can push."[1]

The vision and core values that motivate the formation of these new economy energy cooperatives inspire a new type of cooperative. Traditional cooperatives exist to serve their members' direct needs. The new energy cooperatives exist to create a just and sustainable future for all: to reduce our energy use and to generate clean power and profits that get recycled back into our communities.

New energy cooperatives are providing energy efficiency, rooftop solar, community shared solar, biofuels, renewable heating systems, and other renewable assets. They are also providing access to electricity services for limited-resource communities and communities of color: creating green jobs, developing innovative community education programs, advocating for community-based energy policies, and promoting innovative state incentive programs with a focus on economic, environmental, and social justice.

New energy cooperatives are also pooling their members' buying power and capital to gain protection from the meteoric rises and crashes in the commodity energy markets. They are finding new ways for people to live in harmony with other people and the living Earth. They are overcoming barriers to the new economy—creating new legal/governance structures, raising capital from the people they serve, addressing regulatory challenges, designing new organizing principles, scaling community-benefits access to traditional sources of capital, and more.

This chapter discusses the emergence of this new type of energy cooperative and provides examples of initiatives that are leading the way.

Co-op Basics

Cooperatives are effective tools for returning economic power to people and communities. The laws that govern cooperatives ensure they are democratic (one member, one vote) and that the benefits and profits of a cooperative go to the members, based on how much they use the cooperative's services. For example, if you are a member of a consumer co-op and you purchase $500 a month and another member purchases $100 a month, you'll receive a larger share of the profits.

People can join together to form a co-op to meet their common needs. Because the law honors cooperatives as a way for people to work together, co-ops are not taxed at both the corporate level and the individual level—only at one or the other. For-profit corporations pay tax at both levels—once when they pay their corporate taxes and again when the individuals receive a dividend and pay taxes on that income.

You can understand cooperatives by looking at who owns them.

United in a *consumer-owned* cooperative, consumers can form a buying group to access goods and services on a not-for-profit basis and to provide good jobs for people in their community. They are enfranchised to make key decisions about how their cooperative operates. Primary examples in the United States include food co-ops, housing co-ops, credit unions, and rural electric co-ops, which have provided democratically run alternatives to for-profit grocery chains, housing developments, banks, and utilities for over one hundred years. When consumers own a cooperative, it exists to serve their needs. It operates on a not-for-profit basis because its revenues come from its members and any profits return to its members. The legal structure requires the cooperative to focus on meeting the members' needs.

United in a *worker-owned* cooperative, workers can create and oversee their own jobs, and they are enfranchised to make key decisions about how their cooperative operates. Worker co-ops are present in every sector of the U.S. economy and are a key business structure for giving a voice back to workers and providing a structure for workers to address their own needs.

Cooperatives can also be owned by farmers and others who produce goods and services and who benefit from working with other producers to market, distribute, and add value to what they produce. Co-ops can be owned by government entities, other cooperatives, small businesses, and large businesses.

And some cooperatives are hybrids, with both worker and consumer ownership, for example.

Like any other business structure, the cooperative model is a tool. The results achieved by a cooperative are defined by the owners' vision and values. Like corporations, they buy and sell and participate in a marketplace for their services. Yet, unlike many corporations today, cooperatives are better designed to reflect a community's vision and values—and to maintain those values over time (figure 10-1). They are a tool member-owners can use to expand access to products and resources in their community while building shared wealth. If the member-owners want to use their cooperative to work for economic democracy, they can. There is a growing movement of communities recognizing the potential of cooperatives in building the enterprises that can drive us toward a just energy transition in the coming years.

INTERNATIONAL CO-OP VALUES AND PRINCIPLES

The values and principles that motivate cooperatives around the world were formally stated by the International Cooperative Alliance in 1995, as follows:[2]

> **Definition**: A cooperative is an autonomous association of persons united voluntarily to meet their common economic, social and cultural needs and aspirations through a jointly owned and democratically controlled enterprise.
>
> **Values**: Cooperatives are based on the values of self-help, self-responsibility, democracy, equality, equity and solidarity. In the tradition of their founders, cooperative members believe in the ethical values of honesty, openness, social responsibility and caring for others.
>
> **Principles:**
>   1. Voluntary and Open Membership—Cooperatives are voluntary organizations, open to all persons able to use their services and willing to accept the responsibilities of membership, without gender, social, racial, political or religious discrimination.
>   2. Democratic Member Control—Cooperatives are democratic organizations controlled by their members, who actively participate in setting their policies and making decisions.

Figure 10-1. Energy cooperatives, like other cooperatives, are owned and controlled by their owner-members. Each member has one vote in membership meetings, where key decisions are made that reflect the co-op's vision and values. *Image source: Erik Hoffner*

3. Member Economic Participation—Members contribute equitably to, and democratically control, the capital of their cooperative.

4. Autonomy and Independence—Cooperatives are autonomous, self-help organizations controlled by their members.

5. Education, Training and Information—Cooperatives provide education and training for their members, elected representatives, managers and employees so they can contribute effectively to the development of their cooperatives. They inform the general public—particularly young people and opinion leaders—about the nature and benefits of cooperation.

6. Cooperation among Cooperatives—Cooperatives serve their members most effectively and strengthen the cooperative move-

ment by working together through local, national, regional and international structures.

7. Concern for Community—While focusing on member needs, cooperatives work for the sustainable development of their communities through policies accepted by their members.

COOPERATION WORKS

Cooperative enterprises have almost 1 billion members worldwide, employ 250 million people, and generate $2.2 trillion providing the services and infrastructure society needs to thrive. Nearly 30,000 cooperatives in the United States represent 256 million members.[3]

In 1844, working-class people in England established a consumer-owned food cooperative, called the Rochdale Society of Equitable Pioneers, when they were no longer able to buy basic staples like butter, flour, sugar, oatmeal, and candles. These cooperatives grew out of collective-benefit economies that had existed for centuries. When surrounded by a food desert, they joined together to purchase the food they needed. Within ten years, there were 1,000 cooperative food stores based on the principles established by these first Rochdale Pioneers.

The Catholic Church has supported the development of cooperatives around the world as part of its commitment to human dignity. For example, in the 1920s in Atlantic Canada, Father Moses Coady, a Roman Catholic priest, organized consumer-owned cooperatives that sold food, clothing, and other consumer goods when corporations moved away and left a severely depressed economy. These cooperatives sustained life in communities that would otherwise have not been able to survive.

In her book *Collective Courage*, Jessica Gordon Nembhard describes the challenges African Americans faced as they developed cooperatives to meet the needs of people in their communities.[4] They built consumer-owned cooperatives and producer cooperatives in the early 1900s to sustain their communities when they were not allowed to participate in the mainstream economy. Nembhard reports that W. E. B. Du Bois encouraged African Americans to "form a group economy based on a sense of solidarity and use producer and consumer cooperatives to position ourselves to serve our economic needs separately from the white economy. This way we could control our own goods and services and gain income and wealth—stabilize ourselves and our communities. Then if we wanted to join the mainstream economy, we could join from a position of strength."[5] Du Bois conducted a study of African American economic cooperation in 1907. According to Nembhard, "In addition to the thousands of mutual aid societies, he noted 154 co-ops."

"Given that anyone organizing any form of economic cooperation in the black community was targeted for intimidation and physical harm, these efforts were acts of courage. As soon as co-ops were officially recognized in Europe, after 1844, the U.S. started to embrace co-ops, and African Americans did, too. We started using formal co-ops with some of the early integrated labor unions in the 1880s, and we created our own Colored Farmers Alliance and Cooperative Union in 1886, which also promoted cooperatives and credit unions."[6]

The Grange movement built farmer-owned cooperatives in the mid-1800s to give farmers a way to sustain their farms. Farmer-owned cooperatives now sustain rural communities around the world.

Edward Filene organized a national network of consumer-owned credit unions to give working-class people access to financial services in the early 1900s. The World Council of Credit Unions now includes 49,000 credit unions, serving 177 million members in 96 countries.

Because of their contribution to "socio-economic development, particularly their impact on poverty reduction, employment generation and social integration,"[7] the United Nations declared 2012 as the International Year of Cooperatives. Then UN Secretary-General Ban Ki-moon announced, "Through their distinctive focus on values, cooperatives have proven themselves a resilient and viable business model that can prosper even during difficult times. This success has helped prevent many families and communities from sliding into poverty."[8]

Moreover, Pope Francis preaches that cooperatives create a "new type of economy" that allows "people to grow in all their potential," socially and professionally, as well as in responsibility, hope, and cooperation. He exhorts the cooperative movement to join the global economy to promote both "an economy of honesty" and "a healing economy."

He urges cooperatives to exercise "the courage and the imagination to build a just path, so as to integrate development, justice, and peace in the world."[9] Further, Pope Francis urges that cooperatives "continue to be the motor for lifting up and developing the weakest part of our local communities and of civil society."[10]

In summary, cooperatives are recognized as an essential part of our local economies around the world. They hold great promise for the movement for economic democracy.

The Rise of U.S. Electric Co-ops

Amidst the great depression of the 1930s, President Roosevelt took on the challenge of bringing electricity to rural America. "In 1935, about 10 percent of . . . [U.S.] farm families were receiving central-station electrical service as compared with almost 95 percent in France, 90 percent in Japan, 85 percent in Denmark, and 100 percent in Holland."[11] It became an economic necessity to bring electricity to rural America. However, private utility companies argued that it was too expensive to build poles and wires in areas with few customers.

Between 1935 and 1958, the U.S. government built locally owned energy cooperatives across 75% of the land mass of the United States. In 1935, Roosevelt established the Rural Electrification Administration, which provided technical assistance for rural electric cooperative development. Congress followed a year later with the Rural Electrification Act to provide low-cost loans so these rural electric cooperatives could build their poles and wires, transmission and generating facilities. Twenty-three years later, 90% of U.S. farms had electricity. By lowering the cost of capital and providing cooperative development technical assistance, the federal government was able to successfully address a market failure and bring electricity to millions of Americans. Imagine if this support for cooperative development could be available today.

According to the National Rural Electric Cooperative Association, there are 840 distribution and 65 generation and transmission cooperatives in the United States, serving an estimated 42 million people in 47 states. They provide electricity to more than 12% of the electric meters in the United States.[12] Some of these cooperatives have become part of the energy establishment, with management unresponsive to the climate crisis or the needs of their members. A number of initiatives are under way to make these rural electric cooperatives accountable to their members, as described in chapter 6.

BUILDING A NEW ENERGY COMMONS

The movement for a new kind of energy cooperative interjects a strong value proposition into the cooperative tradition. Cooperatives provide a structure for just governance and principled management that allows them to reflect their members' values. When the members are focused on restoring justice, treating the Earth and all living beings with reverence, learning to live in harmony with others and with the Earth, then the cooperative is an instrument of those values. The cooperative can function as part of a commons,[13] a resource that is fairly shared and stewarded by a community.

Traditional cooperatives are often used to enclose a commons, to keep resources for the benefit of a small group, instead of looking for ways to share those resources fairly for the benefit of the community. For example, housing cooperatives can be used to provide affordable housing for all, or they can be used to legally discriminate against families with children or against people of color. Electric cooperatives can provide clean energy for all, or they can lobby against clean energy legislation and expand into urban areas without allowing those new customers access to membership and democratic participation.

Community-based cooperatives are locally owned and controlled. And, as great as local ownership is, we need to move beyond talking about "ownership" to looking at our "relationships" with one another and the natural world. When we talk about ownership, customers, clients, labor, and resources, we separate ourselves from one another and nature. These are words of colonization, of taking power over others and over nature. New energy cooperatives are looking at a more indigenous understanding of our relationship to other people and the Earth, moving away from "ownership and control" toward relationship, listening, and connection.

The new energy cooperatives described in this chapter are facing these challenges head-on—to bring people together and make a difference, especially in the communities most marginalized in our culture. For most of us, there are more questions than answers. What experience do we have in building a deep spiritual connection with one another and nature? Where can we meet . . . as the descendants of settlers and slave owners and as the descendants of indigenous people and slaves? Can we look unflinchingly at the violence that created the class structures of privilege and disenfranchisement we're living with today? Can we find a healing way to acknowledge the theft of land, of labor, of life that our lives are built on? How do we relate authentically with people we've stolen from? Or who have stolen from us? How do we foster reconciliation and reparations?

How do energy cooperatives, as part of an economic democracy movement, ensure a commitment to restorative justice is built into our DNA? These cooperatives have learned that building projects together works—working shoulder to shoulder, side by side, to bring clean energy and energy efficiency, to bring good green jobs, to bring hope to marginalized communities.

As cooperatives generate revenue, they return that cash to local vendors and employees and pass back the profits to their consumers. This nonextractive finance model helps community-based organizations remain resilient to market shifts and continue to provide stable service and jobs in their communities.

ANATOMY OF THE NEW ENERGY COOPERATIVE

Consumers are organizing now across the country to create a new kind of energy cooperative. The new energy cooperatives are providing different energy products and services as compared with the traditional rural electric cooperatives that owned the poles, wires, generation plants, and transmission lines. At the core, the new energy cooperatives:

- Have a commitment to sustainability and reversing climate change.
- Are organized as a multiclass, multirace movement to address social, economic as well as environmental justice.
- Have members whose primary need is to have a just and sustainable

energy future, so they don't stop at meeting their current members' needs but seek to serve the energy needs of their region (their future members), especially the energy needs of communities of color and communities with limited resources.

- Are committed to local ownership, democracy and self-determination.
- Are Earth reverent.

There are many examples of how communities are building these new energy cooperatives to meet their needs. They are working to build an energy system that:

- Reduces energy waste and energy use;
- Provides affordable, sustainable heating, cooling, light, and power for all;
- Provides good jobs, especially for people in limited-resource communities and communities of color, with costs and profits shared equitably;
- Reduces use of fossil fuels, biomass incineration, and nuclear power.

Example of New Energy Cooperatives

A number of energy cooperatives are taking the lead in defining the new kind of energy cooperative described in the previous section of this chapter. Here we describe the energy cooperatives, Co-op Power, Cooperative Energy Futures, Energy Solidarity Co-op, and Ecopower: what they've been able to accomplish and the opportunities and challenges before them.

CO-OP POWER

In the early 2000s in the northeastern United States, a perfect storm was brewing in which the electric utilities were about to be deregulated, activists were organizing to fight climate change, and entrepreneurs were experimenting with various renewable energy technologies. Consumers had few sustainable energy choices. Solar systems installed in the 1970s sat idle, with few solar companies still in operation able to get them back up and running.

People in the Northeast have a tradition of direct democracy in governing their towns through town meetings, and barn raisings are commonly known as events where people come together to build a barn. Though communities are largely segregated by class and race, there are many places where people come together across class and race and get things done. It's also a part of the United States where climate change is having a noticeable impact, not only on the weather, but on the forests, too.

Co-op Power, launched in 2004, is a multiracial, multiclass cooperative movement.[14] It's a consumer-owned energy cooperative working for a just and sustainable energy future. It is also a decentralized network of community energy cooperatives in New England and New York dedicated to working together as agents of social, economic, and environmental change in their region:

- Boston Metro East CEC (serving the Greater Boston area inside Route 128)
- Worcester Community Energy Action CEC (serving Worcester County, Massachusetts)
- Hampden County CEC (serving Hampden County, Massachusetts, including Springfield and Holyoke)
- Hampshire CEC (serving Hampshire County, Massachusetts, including Northampton and Amherst)
- Franklin CEC (serving Franklin County, Massachusetts, including Greenfield)
- Co-op Power of Southern Vermont CEC (serving Windham County, including Brattleboro)
- New York City CEC (serving Manhattan and the surrounding boroughs)

Co-op Power is owned by 550 families and a handful of schools, universities, cooperatives, and small businesses (figure 10-2). The community energy cooperatives in the Co-op Power network all share one cooperative structure to make collaboration easy and reduce administrative costs.

Figure 10-2. Co-op Power regional gatherings give members an opportunity to build relationships in our multiclass, multiracial movement that spans urban and rural communities.
Image source: Erik Hoffner

When members join their community energy cooperative, they join both
their community energy cooperative and Co-op Power.

Altogether, over the last twelve years, community energy cooperatives have invested in developing 24 new energy enterprises to meet their
community's energy needs, including the needs of people in limited-
resource communities and communities of color. They've brought in
income from membership fees and grants, and they've lent money to
build various enterprises providing services members have wanted in
their communities.

Here is Co-op Power's recipe for democratizing energy: First, people
come together across class and race to make change in their community
by using their power as investors, workers, consumers, and citizens ready
to take action together. Then, they work together to build community-

owned enterprises with local capital and local jobs to serve local energy needs. It's a proven strategy for making a real difference.

Below are snapshots of four of the enterprises launched by Co-op Power: Northeast Biodiesel, Energía, Pioneer Valley Photovoltaics (PV Squared), and Resonant Energy.

Northeast Biodiesel Company, LLC

Franklin and Hampshire Community Energy Co-op members decided to build a biodiesel plant to recycle cooking oil as their first project because they believed this business would require the least capital, take the shortest time, and produce profits that could be used to start up other community-based sustainable energy businesses. Despite the best expert advice, things didn't turn out as initially envisioned.

Co-op Power has had to jump through many hoops to secure financing for the $4.5 million project. It also suffered from public relations challenges, consumer confusion, an extended development process, permitting delays, and so forth.

After fourteen years of development, Northeast Biodiesel is launching in the fall of 2017, making 1.75 million gallons of biodiesel from used cooking oil.[15] It will provide biodiesel for diesel engines and oil heat systems. In addition to fourteen jobs at the plant, where workers will own a 25% share, Northeast Biodiesel will create jobs for collecting, processing, and distributing the used cooking oil as well as delivering biodiesel to distributors and Co-op Power members.

Energía, LLC

Hampden Community Energy Co-op members wanted to create good green jobs for people in their limited-resource communities and communities of color. Energía was born in 2009, following a successful U.S. Department of Health and Human Services (HHS) grant application.[16] Co-op Power partnered with community organizations to run a training program using HHS WorkForce Development funds so that young people from the local communities (figure 10-3) would be able to be successful in jobs in this new business.

Figure 10-3. Young people from the local community air-seal a home in Boston, Massachu-setts. *Image source: Lynn Benander*

Eight years later, Energía has $2.5 million a year in revenues providing residential and commercial air sealing and insulation services. The 24 to 28 workers—75% of whom are Latino and 75% of whom are under the age of 25—have formed a worker cooperative that now owns 30% of the business.

Pioneer Valley Photovoltaics (PV Squared)

PV Squared is a worker-owned solar installation cooperative located in Greenfield, Massachusetts.[17] It's been in business since 2002, installing solar electric energy systems for homeowners and businesses in western Massachusetts.

Leaders from Co-op Power worked with community volunteers and solar experts to craft a business plan and secure a grant to launch the cooperative. Co-op Power brought in the first customers ready for instal-lations. What began with four employees has now grown to more than twenty. They have installed PV systems for hundreds of clients and play a leadership role in developing solar policy that serves people and com-munities in Massachusetts.

As a worker-owned cooperative, the business is run by the people who work in it. Workers at PV Squared invest money when they join the business as owners. At the end of each year, they are paid a portion of the money the business makes after expenses.

Resonant Energy, LLC

Resonant Energy was founded in the summer of 2016 to launch the Solar Access Program—a new rooftop solar hosting model designed for the 50% of households, nonprofits, and other institutions that lack the capital or creditworthiness to qualify for a loan or a traditional power purchase agreement.[18] As a community-based solar development company, Resonant Energy employs innovative financing strategies to leverage the buying power of local anchor institutions—including businesses, nonprofits and municipalities—to reduce the credit risk of putting solar on low-income households and to dramatically expand urban solar access, jobs, and education.

COOPERATIVE ENERGY FUTURES

In a city that gets an average annual snowfall of 32 inches and temperatures below 50°F for six months of the year, energy costs are a big deal. Cooperative Energy Futures (CEF) is a for-profit, community-owned clean energy cooperative based in South Minneapolis, serving members across Minnesota with their solar projects and energy efficiency services.[19] General Manager Timothy DenHerder-Thomas has been a part of CEF since the beginning and is leading the co-op in its current expansion.

Minneapolis is a thriving midwestern city that routinely makes top-ten lists for best places to live, to find a job, to be healthy, to be a foodie, to live without a car, and to retire. It has a diverse population made up of communities of African Americans, Hmong, Mexicans, Ecuadorians, Indians, Koreans, and German-Scandinavian descendants. It is the second-largest economic center in the Midwest after Chicago and is home to major companies based in health care, commerce, and finance. All the strengths of the metropolitan area make it ripe for opportunity in community building and innovation.

Cooperative Energy Futures, incorporated in early 2009 as a 308B cooperative, started off lean, focusing its energies on providing members with energy efficiency workshops, and bought supplies with funds from a small grant. Another low-capital initiative they undertook was doing member recruitment through solar site assessments and managing the solar sales process for members. Timothy credits these kinds of projects for building long-term grassroots relationships with local leaders and residents in South Minneapolis. Many residents in these communities were also struggling with high energy bills and economic instability.

Cooperative Energy Futures does group contracting for deep insulation retrofits to improve existing building stock. Education is also an opportunity they have tapped into through do-it-yourself trainings for weather stripping, caulking, and home energy systems. More recently, after Minnesota adopted a community solar law, CEF has shifted to developing community shared solar projects; bulk buying insulation and residential solar were reaching only a small portion of their membership.

The cooperative accomplished a lot in its first seven years. Early on, it completed the first couple of rounds of insulation bulk buying for twenty-five insulation installs in the city. They brought quality control and expertise for members as project consultants. Between 2014 and 2015, CEF installed 7.5% of all residential solar that received Made in Minnesota (MiM) state incentives.

CEF has completed a subscription drive for two solar garden projects that, starting in 2017, will provide 110 households with their full power needs for the next twenty-five years. One of their solar garden projects is with a black church in a Minneapolis low-income community of color where project subscribers live in the community. The second solar garden is in the city of Edina, Minnesota, a largely white, middle- and upper-middle-class community. The City of Edina specified it to be for "Edina residents only" as a priority. Multicultural spaces happen on a project basis. Some groups come with that purpose and vision while others are focused on a different set of benefits.

Cooperative Energy Futures operates with universal principles. For example; it does not discriminate by credit score. The CEF solar gardens are accessible in a way that private corporate projects are not. CEF has also partnered with Renewable Energy Partners to do direct job training for residents of northern Minneapolis and require installers to employ people of color as 50% of their labor force, including graduates from the training program.

Reflecting on the challenges that CEF faces, DenHerder-Thomas says, "In the current policy framework, there are a lot of barriers to getting involved. We have a monopoly utility system that constrains our ability to do things outside of the box."[20] Another challenging area has been financing the projects members want. Community solar finance is complicated and requires significant investment from tax equity investors such as banking institutions. CEF works with investors to create a new norm of investment and is trying to get all people to rethink how to create value without extracting wealth from communities.

CEF has sought to create opportunities through clean energy to empower people and make the economy work for everyone. It has had a strong focus on economic access and is interested in job training and economic development in communities of color. CEF empowers communities to create their own future by working together to build, own, use, and manage basic energy systems that sustain the community.

ENERGY SOLIDARITY CO-OP

The West Coast provides an interesting opportunity for renewable energy cooperatives since many energy policies are decided on a state-by-state basis. Energy Solidarity Cooperative (ESC) is recent start-up in Oakland, California, that designs and builds community-driven, cooperatively owned solar energy projects and political educational programs.[21] They focus on building relationships with such groups as community-based organizations, schools, and places of worship in communities of color and with low-income residents.

Oakland has long been a working-class city in the San Francisco Bay Area. Being the eighth-largest city in California, Oakland serves as a trade center for San Francisco. It saw demographic changes between 2006 and 2016 but is still very diverse, with 65.5% of residents being nonwhite. There are long-standing African American, Hispanic American, Asian American, and Native American communities as well as recent immigrants from China, Vietnam, the Philippines, Southeast Asia, and Central and South America.

The cooperative sees renewable energy as the opportunity to build community power—social, political, infrastructure, economic, and genera-

tive. They see this as being part of the just transition toward solidarity economies. The cooperative is comprised of worker-owners, consumer-owners,
and community-driven funders—an innovative approach that works to
bring the often differing interests of each group under one umbrella.

These stakeholders work to build out more community-owned clean
energy systems and energy efficiency measures in underserved communities. They finance energy systems and efficiency measures in a way
that keeps money in their community and creates a more secure path to
refinancing once they are creditworthy. They work to save revenue for
organizations operating at underserved sites while also generating revenue for worker services. Also essential is their focus on growing the clean
energy cooperative workforce with ecological and climate justice educational and training programs.

ESC worker-owner Ashoka Finley is focused on building out
community-owned rooftop solar and sees his work as being deeper than
technology problem solving. "It would be so much easier if we just developed solar projects and didn't build relationships—we would just pick
the best sites and that would be it. But we're focusing on job growth and
education, which can hinder the project's development in some ways."[22]

Energy Solidarity Cooperative offers several products and services to
meet its mission. Energy efficiency is usually a great first step so ESC conducts
residential or institutional preliminary energy audits and minor building
retrofits. Buying solar can be a difficult process for homeowners to navigate,
so ESC assesses solar potential and site selection for renewable energy systems in neighborhoods. The Co-op also assists with the development of management models, installation partnerships, and innovative financing.

EUROPE'S ENERGY CO-OPS

The roots of the modern renewable energy cooperatives in Europe began in
the 1970s in Denmark with the development of wind power.[23] Germany was
the next biggest player to get into renewable energy co-ops in the early 1990s,
also with wind power. In the United Kingdom, a citizen group's work on
broad community ownership resulted in operating wind turbines in 1995.

Deregulation across Europe has been a great opportunity for social and cooperative entrepreneurs to organize alternative models to monopolies for renewable energy and community ownership of energy resources. Many of the movements were spawned as a way to both disrupt monopolistic utilities and reduce over-reliance on nuclear power. Renewable energy has also seen development in corporate models, as in Spain, where early corporate wind development actually made it harder for cooperatives to become established.

In 2011, REScoop (Renewable Energy Source Cooperative) was founded to help co-ops across the continent share resources and best practices and manage policy. As of April 2016, it included twenty member organizations in eleven European Union states. Citizens can join and form their own REScoops as a way to develop citizen-owned renewable energy projects and are provided with resources to get set up. REScoop holds that energy democracy has a value and views the energy transition as an opportunity to move from fossil fuels and nuclear to renewable energy, from monopoly to citizen ownership, and from people being consumers to being "prosumers" by using the energy that they themselves produce.[24]

One of the most successful energy cooperatives in Europe is Ecopower, a Belgian consumer cooperative founded in 1991 that has built its membership to more than 45,000 and is currently supplying green power from hydro, wind, and solar energy generation. It has also provided support services to help members with insulation of homes, providing building analyses at the front end and checking work quality of contractors at the back end.

Like the U.S. cooperatives discussed earlier, the Ecopower cooperative demonstrates the power of creating relationships with local municipalities, collaborating with other cooperatives, and building membership as powerful stepping stones. These basic practices have helped Ecopower remain competitive with private corporations, although it is facing new challenges with respect to electricity pricing and corporate efforts to co-opt co-op messaging.

Dirk Vansintjan, one of Ecopower's founding board members, sees collaboration between energy cooperatives becoming more important. He is working across Europe to build a revolving fund that can help cooperatives across the continent overcome their financing hurdles.

Toward A Cooperative Energy Future

As we struggle to address climate change, it is clear that the technology exists to move us off of fossil fuels and that the structures exist to move us to a new economy. In the energy industry, we have one of the most significant opportunities for cooperative development ahead of us. The United States and the rest of the world are poised for a trillion-dollar investment in renewable energy, grid upgrades, electricity expansion, and distributed energy resource deployment to power the twenty-first-century economy.

This is an energy future in which the same corporations that have brought us to the brink of climate disaster will be first in line to profit from the clean energy revolution, and environmental justice communities spawned in the wake of the fossil fuel economy will be the last to see clean and reliable energy in their neighborhoods. Cooperatives offer a vehicle for changing that dynamic, at least for some—an alternate future in which community leaders democratize business ownership, seed local investment, share resources, maintain fair pricing, and organize consumer power in a way that not only reduces costs, but also takes political power. Cooperatives have a key role in the fight for community resilience.

Will the energy industry continue to draw all of the wealth and power to the wealthiest 1% on the planet, or will the 99% rise up and take control as advances in distributed renewable energy create opportunities for a more community-based energy economy to flourish? Decentralized power makes it possible for people to steward energy resources. Communities can use cooperatives to build power, meet their needs, and share resources.

When people come together in a cooperative and commit to educating their communities about our energy future, they wrestle with the consequences of the extraordinary waste in energy production and transmission, industrial agriculture, waste processing itself, transportation systems, and so forth, making it clear that the real solution to our energy challenges is to shift away from an extractive economy, away from the materialism that generates so much to throw away, away from using up the Earth and filling it with trash, and instead toward a more loving, equitable, just, ethical, gentle, connected, and Earth-reverent way of living together on the planet.

More than half of our U.S. economy is comprised of small businesses, nonprofits, cooperatives, and government entities, yet most of our economic development resources go to support large corporations.[25] Cooperatives can provide jobs and products and services for our communities. Real power and strength come from communities and from cooperation.

Communities across the United States are successfully addressing these challenges and demonstrating a pathway forward with committed groups of activists and entrepreneurs. The pathway for each cooperative has been very similar. An ambitious and motivated group involves people in the community to envision how they want to own and develop energy resources. Once a vision is laid out, the members pursue a winnable project. An early win helps to build momentum for membership building, after which the next target is a larger project to involve even more people. Community organizing is essential with like-minded nonprofits, student groups, churches, and other cooperatives working together to see specific policies changed or implemented to support new economy project development. It is important to have a diverse and committed board of directors who bring various business development, engineering, energy, legal, finance, and community organizing expertise. The other ingredients for hanging in for the long haul is to have a clear governing structure as well as making time to celebrate all wins, big and small.

Co-ops build community resilience as energy commodity markets and venture-funded companies rise and fall around them—and they offer a legitimate pathway to serving communities who have been systematically barred from wealth-generating opportunities, jobs, and political enfranchisement in our current economic system. New economy energy co-ops are not just a good tool—they're a cornerstone of a new energy economy.

1. Richard Heinberg, author of *Our Renewable Future: Laying the Path for One Hundred Percent Clean Energy*, coauthored with David Fridley (2016), interview by Lynn Benander, January 19, 2017.
2. International Cooperative Alliance, "Statement of Co-operative Identity, Values and Principles," http://ica.coop/en/whats-co-op/co-operative-identity-values-principles.

3. International Cooperative Alliance, "Facts and Figures," http://ica.coop/en/facts-and-figures.
4. Jessica Gordon Nembhard, *Collective Courage* (Pennsylvania State University Press, 2014).
5. Jessica Gordon Nembhard, in an interview with Ajowa Nzinga Ifateyo, June 13, 2014, published in Grassroots Economic Organizing.
6. Ibid.
7. UN Secretary-General Ban Ki-moon, United Nations declaration. "2012, the International Year of Cooperatives," http://www.un.org/en/events/coopsyear.
8. Ibid.
9. Pope Francis, "Pope Urges Co-ops to Promote Economy of Honesty," *Radio Vatican News*, February 2, 2015, http://en.radiovaticana.va/news/2015/02/28/pope_urges_co-ops_to _promote_economy_of_honesty/1126264.
10. Christopher Dodson, editorial, North Dakota Catholic Conference, "The Co-op Pope," November 2015, http://ndcatholic.org/editorials/column1115/index.html.
11. Robert T. Beall, "Rural Electrification," *1940 Yearbook of Agriculture* (USDA, US Printing Office), p. 790.
12. National Rural Electric Cooperative Association, "Co-op Facts and Figures," http://www.nreca.coop/about-electric-cooperatives/co-op-facts-figures.
13. David Bolier, *Think Like a Commoner* (New Society Publishers, 2014).
14. Co-op Power, htttp://www.cooppower.coop.
15. Northeast Biodiesel Company, LLC, htttp://www.northeastbiodiesel.com.
16. Energía, htttp://www.energiaus.com.
17. PV Squared, http://pvsquared.coop.
18. Resonant Energy, http:// www.resonant.energy.
19. Cooperative Energy Future, https://cooperativeenergyfutures.com.
20. Timothy DenHerder-Thomas, phone interview with Diego Angarita Horowitz on October 31, 2016.
21. Energy Solidarity Co-op, htttp://www.energy-coop.com.
22. Ashoka Finley, interview with Renewable Cities, CA, June 2, 2016.
23. J. H. M Larsen, et al., "Experiences from Middelgrunden 40 MW Offshore Wind Farm," Copenhagen Offshore Wind Conference, DK, October 26–28, 2005.
24. European Federation of Renewable Energy Cooperatives, htttp://REScoop.eu.
25. Michael Shuman, *Local Dollars, Local* Sense (New York: Chelsea Green Publishing, 2012).

CHAPTER 11

Building Power Through Community-Based Project Development

ANYA SCHOOLMAN AND BEN DELMAN

Solar energy offers many advantages over traditional electricity sources. The fuel is free and produces zero emissions. Solar power can be deployed at a small scale, and to the benefit of individuals and communities. This cannot be said about most other energy sources. Just try putting a natural gas plant on your roof! Plus, in most of the country, the cost of solar power is now lower than, or comparable to, electricity from fossil fuels.

The potential for solar power to give individual citizens greater control over their energy future (and bills) is unquestionably large. However, can we deploy small solar arrays at a large enough scale to increase equity at the neighborhood or city level? One answer to this challenge comes from Washington, DC—not the DC of lobbyists or blue-ribbon commissions, but the DC of working families and long-time residents.

Since 2009, a solar revolution has been growing in the city's neighborhoods. Away from politics and partisan rhetoric, groups of DC neighbors have come together to make solar energy a reality in their communities. Their efforts have succeeded by connecting people to a critical need, energy autonomy, via solar energy. Their work has contributed to a national movement for locally owned renewable energy projects. Importantly, these DC residents have also banded together to drive policy reform to help more people go solar. In doing so, they have created a virtuous cycle of project implementation and policy change.

This chapter tells the story of building a solar bulk-purchasing co-op movement, using its base of support to push for the broader democratization of energy, and the implications of that effort for defining the electricity system of the future.

A New Model for Growing Clean Energy Emerges in DC

In 2007, two twelve-year-olds, Walter Lynn and Diego Arene-Morley, asked their parents to put solar on their homes in Washington, DC's Mt. Pleasant neighborhood. The boys were inspired to go solar by the movie *An Inconvenient Truth* as a way to combat climate change. Walter's mother, Anya Schoolman, made several phone calls to local installers and discovered that going solar was an expensive and complicated process. Like all hardheaded preteens, Walter and Diego would not be talked out of the idea. So, Schoolman wondered if some sort of bulk purchase might make solar affordable.

Within two weeks, Walter and Diego had engaged fifty other neighbors who wanted solar on their roofs. Many of the homeowners just wanted to gain control of their rising energy bills. Others wanted to take action on climate change. Many were motivated by the idea of freeing themselves from the local monopoly utility. Thus was born the city's first solar co-op, the Mt. Pleasant Solar Cooperative. The co-op worked tirelessly for two years, meeting in living rooms, getting help from local solar providers, and sharing their expertise and knowledge with one another. In September 2009, the co-op celebrated its forty-fifth solar installation in Mt. Pleasant, and by the 2012 Mt. Pleasant Solar Fair, Walter and Diego had become local heroes for catalyzing a new energy movement (figure 11-1). From there, the group expanded to help neighbors form co-ops across the city, creating DC Solar United Neighborhoods (DC SUN).

In 2011, Anya Schoolman founded a national nonprofit, Community Power Network (CPN) to disseminate and replicate the successful project model DC SUN had developed. Today, in addition to DC SUN, CPN now runs Solar United Neighborhoods programs in the states of Florida (FL SUN); Maryland (MD SUN); Ohio (OH SUN); Virginia (VA SUN); and West Virginia (WV SUN). CPN also works to connect organizations and

Figure 11-1. Diego Arene-Morley (left) and Walter Lynn (right) interviewed at the 2012 Mt. Pleasant Solar Fair *Image source: Community Power Network*

community groups across the country that are creating and experimenting with other innovative ideas for community-driven renewable energy. CPN helps communities start their own solar co-ops, protect their right to produce solar power, and implement policies and project models that expand solar access to low-income households. Schoolman found that communities across the country, from low-income, inner-city neighborhoods to conservative rural areas, were clamoring for the same thing: to take control of their own energy future.

DEVELOPING THE SOLAR CO-OP MODEL

The Mt. Pleasant Solar Cooperative grew out of the interests and needs of a group of neighbors, and it could only succeed with a committed community behind it. Anya, Walter, and Diego quickly discovered that the level of interest in solar and commitment to the group was even greater than they had hoped.

Building an energy co-op movement is an organic process, growing from individual to community to citywide change (figure 11-2). In the beginning, the most important thing was for group members to clearly identify their own goals: Why are we doing this? What is important to us? Do we want to lower our electric bills? Do we want to create local jobs? Do we want to become entrepreneurs in our community? The group decided that its goal was to take as many houses solar as possible at the most economical price. Additional priorities were to reduce electric bills and jump-start local solar businesses, enabling them to scale up and expand the market across the District. Identifying goals is an essential first step of any collaborative process. Once the goal (in this case, helping as many people as possible go solar) is clear, the project model can follow.

The Mt. Pleasant Solar Cooperative next identified their baseline understanding of solar and educated themselves to ensure they would make an informed decision. The co-op researched, held meetings, and found experts to help them understand solar technology, costs, financing options, and policies. Each group member brought particular skills and perspectives to the process: from the roofer and the electrician to the website builder and the meeting organizer, everyone had something to offer. One member contributed cakes baked in a solar oven to the group's regular meetings. Sharing and welcoming each member's diversity of experience made the group more effective and had a profound impact on the group's ability to see the project through to success.

Educating themselves about solar was critical but turned out to be only one piece of the puzzle. As the co-op grew, members realized they would need to fight for stronger solar policies if they wanted to help more of their neighbors go solar. The rules governing energy markets are, in large part, stacked firmly against democratizing energy. In many states, energy companies, especially large, vertically integrated, investor-owned utilities (IOUs), play a dominant role in developing and writing energy rules. These companies work hard to ensure that regulators support their desired outcomes. The utilities spend millions in campaign contributions and lobbying each year. This enables them to make persuasive arguments to friendly regulators. They also benefit from the revolving door between

Go solar.

Join together.

**Fight for
our energy rights.**

Figure 11-2. Community Power Network model of energy transformation: from solar installation to solar community to citywide solar policies *Image source: Community Power Network*

regulatory agencies and the industry. This state of affairs has created a lopsided market that supports monopoly utilities and a regulatory system that reinforces the status quo.

Decentralized, democratized green energy is not yet part of the mainstream.

FROM PROJECT TO POLICY

As far back as in 2007, the District had an aggressive renewable portfolio standard (RPS) with a "solar carve-out."[1] The RPS and solar carve-out, in theory, required the local utility to obtain a certain percentage of its electricity from solar sources. In reality, this was not happening. The legislation authorizing the RPS set the alternative compliance payment (ACP) for utilities too low. The ACP is the penalty a utility must pay if it does not meet the RPS requirements. Because the ACP was so low, the local utility was paying the ACP rather than procuring more renewable energy.

As the local activists in the Mt. Pleasant Solar Cooperative searched for a way to make solar accessible and affordable to all their members, the ACP problem began to stand out. "Why is the solar market in DC so small?" they wondered. Based on the RPS, much more solar should have been getting built. The ACP loophole turned out to be the biggest problem. Without an enforceable requirement to put solar on the grid, the utility had no incentive to do so.

In 2008, the Mt. Pleasant Solar Cooperative and its allies secured legislation to double the District's ACP. Raising the penalty for not having enough solar-generated electricity encouraged the utilities to buy solar renewable energy credits (SRECs)[2] from solar producers. This demand drove up the value of SRECs, making solar electricity production more financially attractive, which, in turn, spurred a huge boom in DC's market for decentralized solar.

However, shortly thereafter, because of an influx of out-of-state projects into the District's market, and a resulting oversupply of SRECs, the value of SRECs crashed, undermining local investments in solar. So activists worked on another piece of legislation to double the solar carve-out requirement (increasing demand for SRECs) and close the market to out-of-state projects (decreasing supply of SRECs). They argued that since DC ratepayers were paying for utility procurement of SRECs via their utility bills, all the jobs and economic benefits should stay in the District—and they were successful in limiting out-of-state projects.

Along the way, there were many other fights as well. Solar activists convinced the local utility to improve its interconnection process, which enables a customer's solar system to connect to the grid. These activists also continue to fight, multiple times per year, to ensure that the grant money appropriated to the District government to develop solar projects for low-income households is spent effectively.

Between 2013 and 2015, the District government installed solar for free on roughly 130 low-income homes per year. In July 2016, activists successfully passed legislation to create a Solar for All program, which increased DC's investment in low-income solar from $1 million to more than $20 million annually. The goal of Solar for All is to reduce the bills of 100,000 low- and moderate-income households by 50% by 2032 using rooftop solar. The program also aims to transform the solar market and create thousands of local green jobs over the next fifteen years.

Each time local solar advocates have taken on a new fight, it has been in direct response to barriers they identified in attempting to install solar or to implement a project to make solar affordable and accessible to all residents. DC SUN activists have successfully repeated this model—linking policy advocacy to on-the-ground projects—numerous times since 2007. With each cycle of projects and advocacy the movement has grown broader, stronger, and more diverse. Through this work, solar advocates have been able to reshape the market as well as the attitudes of activists, politicians, and homeowners. One lesson for other movements is that it can be more effective to start with a small, doable, impactful project and build, expand, diversify, and scale from there.

A decade after the Mt. Pleasant Solar Cooperative formed, the District of Columbia is now one of the most solar-friendly cities in America, home to one of the hottest solar markets in the country. Most important, the market is supported by activists from all parts of the city, not merely in the upper-class Northwest quadrant, but also in Anacostia, in Southwest DC, and Brookland, in Northeast DC. With the landmark Solar for All program, DC has the potential to show the rest of the country how solar can truly transform energy markets to benefit all its residents.

The CPN Co-Op Model

Community Power Network (CPN), the national initiative that grew out of DC Solar United Neighborhoods (DC SUN), built on these early experiences working with neighbors across DC to refine its community-led solar co-op model. Solar co-ops today are similar to the Mt. Pleasant project, but the process has become much more efficient, professional, and streamlined. What took the Mt. Pleasant Solar Cooperative two years can now be completed in nine months or less.

The central idea of a neighborhood solar purchase co-op is to enable a group of neighbors to go solar together and get a bulk discount, thereby making solar more affordable and accessible. By going solar as a group, participants save on the cost of their system and get support from their peers as they go through the process. The process is still democratically run by the local community, but CPN provides professional technical assistance so that each group does not have to reinvent the wheel and can benefit from the experience of previous co-ops.

As the local group goes through the solar co-op process, it learns about solar technology, installation, financing options, and policies that impact its ability to go solar. Common barriers include restrictive fire codes and homeowners association rules; net metering limitations; weak or nonexistent RPS standards; and cumbersome permitting, inspections, or interconnection practices. Through engagement in the co-op process, people come to understand how energy policy is made and what kinds of actions are needed to change it. They are motivated to take action because these barriers directly impact their ability to put solar on their own homes.

CPN'S FOUR-PHASE PROCESS

The co-op model Community Power Network has developed is a four-phase process. In the first phase, CPN works with a local community partner to spread the word about the co-op, create excitement about solar, and recruit residents to attend information sessions. This local community partner can be a nonprofit, a church, a local government agency, or simply a group of individuals. At information sessions, CPN explains the co-op process, solar technology and installation, financing, and policy issues, in addition to answering the many detailed questions people ask about solar. Interested homeowners can then join the co-op online. After a participants sign up, CPN does a preliminary screening of their roofs and properties via satellite imagery to ensure they are good fits for solar and sends them additional information about solar.

The second phase begins once a critical mass of homeowners (usually twenty to thirty) have signed up and passed the roof screening. CPN, with input from the co-op members, issues a request for proposals (RFP) seeking bids from area installers. Each community has the opportunity to customize the RFP to reflect local values and preferences. For example, if the community wants to give priority to a local company, that will be expressed in the RFP. Once CPN has received bids from installers, it develops a detailed analysis of the bids and helps the co-op establish a selection committee. The members of this committee review the bids and select an installer to complete all the solar projects for the group. While CPN provides guidance and expertise, the decision of which installer to choose is made by the co-op members. Local community values guide the process throughout.

The third phase of the solar co-op begins once an installer is selected. The installer visits all of the co-op members' homes and provides each of them with an individualized proposal for a solar array. The price quote reflects the group discount offered in the winning bid. Participants who decide to move forward with the installation then sign a contract directly with the installer to purchase or lease their solar system. The co-op remains open to new members for about a month after the installer is selected. This provides the local community partner an opportunity to continue to grow the group after the installer has been selected.

CPN works closely with the installer and individual participants throughout the site assessment and contracting phase to provide oversight, ensure good customer service, and move the process forward. This engagement is essential to ensure the chosen installer is staying on schedule and following up with participants in a timely manner. Because CPN builds a high level of trust with the community from the beginning of the co-op, CPN's continued engagement and support help encourage more participants to move forward with their solar projects.

In the fourth phase, co-op participants who signed contracts have their solar systems installed, after which the co-op holds a celebration of their new homegrown clean energy. CPN provides ongoing support to ensure homeowners have a positive experience during installation and the concurrent inspection and interconnection processes. Additionally, CPN provides several community-building resources for co-op members and the broader solar community, including a listserv for each state. The listserv is an online community that allows geographically dispersed solar producers and advocates to connect with one another, answer one another's questions, and discuss developments in solar technology or policy. These platforms connect co-op participants to the growing network of solar supporters in their state and enable people to stay engaged and receive ongoing support.

In this way, these communities serve as a base for ongoing policy actions. Co-op members become invested in the solar movement, build relationships with other solar advocates, and are often excited to take action on policy issues that impact their solar ownership.

EXPANDING SOLAR TO RURAL COMMUNITIES

The success of the purchasing co-op model in the District of Columbia raises the question, can this model work in areas that are less wealthy or more conservative? Absolutely.

More than 100 miles from Washington, DC, the town of Harrisonburg, Virginia (population 50,000) sits in the state's Shenandoah Valley. Harrisonburg is the base of one of the area's largest and most engaged solar co-ops, despite a state policy environment that does not encourage residential solar. Between 2014 and 2016, solar co-ops helped nearly one hundred homeowners in Harrisonburg go solar.

The group's success can be traced to a strong community of activists who wanted to help their neighbors go solar. Members actively spread the word about the co-op to friends and family and worked with local media outlets to broadcast news about the co-op. This helped the group overcome the challenge presented by the co-op's wide geographic territory and to turn it to their advantage.

The area's rural geography proved to be an asset. The prevalence of open space gave many co-op members the option to have a ground-mounted system rather than a rooftop system (rooftop systems are generally smaller and may be more constrained than ground-mounted solar). This flexibility increased the overall number of people able to go solar, especially for those whose roofs do not receive a lot of sunlight.

Leveraging Co-op Communities

Strengthening the community of solar supporters is not simply a means to increase solar adoption, it is a base for defending solar programs. In 2015, West Virginia utilities sought legislative action to roll back the state's net metering policy. This would have strangled the state's nascent solar market. Just as it was finally becoming cheaper to generate solar on residential property than to buy it from the utility, the utilities stepped in to limit West Virginians' ability to produce their own power.

West Virginia's network of solar co-ops quickly sprang into action. They pulled together a legislative steering committee, identified key players in the legislature, developed talking points and action alerts, wrote op-eds, and urged people to share these materials with their friends, families, neighbors, and other networks.

The backbone of this effort was a statewide listserv of solar supporters that Community Power Network developed through West Virginia Solar United Neighborhoods (WV SUN). This group generated more than 600 letters to state legislators in just one week, urging them to oppose the utility-backed legislation. In total, the WV SUN listserv generated more than 1,000 constituent contacts to legislators. The effort bore fruit. Solar activists derailed the worst parts of the anti–net metering bill. In the end, the legislature passed a bill directing the West Virginia Public Service Commission (PSC) to study net metering. In his signing statement, the governor emphasized the increasing role that solar is playing in the state and acknowledged the importance of new solar jobs—a huge change in messaging for a West Virginia elected official.

This moment marked the first time that West Virginia's solar community had come together to speak with one voice. This could not have happened without the foundation of community-building work already laid by solar co-ops across West Virginia.

COMMUNITY SOLAR IN DC

An improving solar market for homeowners has created an opportunity for solar activists to push for policies that would benefit communities previously unable to access solar energy. Beginning in 2011, activists in DC initiated a fight to enable "community shared solar" in the District. This program enables ratepayers, including renters and apartment dwellers, to purchase "shares" in a solar installation and receive a credit on their bill for their shares' electricity production, just as if the solar installation were on their own property. Community shared solar would open solar access to the 60% of DC residents who do not own their home or who live in multifamily buildings.

This community solar effort succeeded in 2013, when advocates urged the DC Council to enact the Community Renewable Energy Act (figure 11-3). This bill enables community shared solar, also known as a virtual net metering program.

Implementation of the law has been a challenge. Against the intentions of the unanimous 2013 Council vote, the DC Public Service Commission (PSC) set the rate for crediting community solar shares at half the retail electricity rate (the rate almost all residents pay), undermining the viability of the program. This low rate discouraged developers from starting any community solar projects in the District, as better project development opportunities existed elsewhere.

Rather than accepting this as a loss, DC activists rallied to put new legislation before the DC Council. Finally, in July 2016, the Council passed a fix to restore the value of community solar shares and ensure that shareholders are treated the same as net metering customers. Once again, the DC PSC slow-walked the implementation, but community solar with *full on-bill credit* for the energy produced came into effect at the very end of December 2016.

This fight illustrates the arduous path we sometimes must take to policies that open access and opportunity for everyone. A broad, robust, and tenacious coalition is needed to take on entrenched utility interests. However, a democratic and equitable clean energy transition is possible if we stay focused on the endgame.

Figure 11-3. **Community residents won the 2013 DC Council vote for community shared solar legislation.** *Image source: Community Power Network*

EXTENDING COMMUNITY SOLAR TO MARYLAND

Maryland's path to community solar followed a similarly winding path. In 2008, several homeowners in University Park, Maryland, tried to go solar but discovered that their roofs were too shaded and thus were not able to move forward. Rather than give up, the neighbors came together and developed their own community solar group. Community Power Network provided technical assistance, recommending that the group create their own special-purpose limited liability company (LLC) to collectively own a solar array on a local church. Their successful project, called the University Park Solar LLC,[3] was copied by several others in Maryland who were impatient to see the democratic energy revolution start. However, the movement was hampered by federal regulations, such as those which place limits on investor solicitation for projects, and utility regulations that make homegrown community solar expensive and complicated. Eventually, the Maryland groups came together to fight for legislation that would enable community solar in the state. After years of lobbying in Annapolis, legislation passed in 2015 that will launch a 196 Maryland community solar pilot program. The first projects are expected to be built in 2017. Many have innovative provisions to ensure that low-income families benefit from the program.

Once the bill passed, the state's solar community worked to ensure that Maryland regulators followed the law's intent—to expand solar access to all Marylanders—when writing the implementing regulations. Solar activists, led by CPN's Maryland Solar United Neighborhoods (MD SUN) program, sent letters to the Maryland Public Service Commission (PSC), urging regulators to set a fair credit rate for community solar projects (a lesson learned from DC's experience).[4] Their success ensured that projects would be developed in the state during the pilot program's time frame. MD SUN also successfully led the fight to make sure that a significant portion of the program was reserved for low-income communities.

ENGAGING THE AFFORDABLE-HOUSING COMMUNITY

Community solar is just one tool solar activists have used to expand the benefits of solar to marginalized communities. Solar activists in Washington, DC have played a key role in pushing for programs to expand solar access to low-income residents and affordable-housing tenants, and in connecting people to these initiatives once they have been created. Activists helped create a sustainable energy utility (SEU), which provides zero- and low-cost energy efficiency and solar installations to permanently lower the bills of low-income households.

DC solar activists also organized a series of stakeholder meetings around low-income solar that for the first time brought together a diverse array of local representatives: federal and local government agencies, affordable-housing providers, economic development organizations, finance and lending institutions, and the solar industry. Solar energy has the potential to be especially beneficial in low-income housing applications, by reducing residents' energy costs. However, the puzzle of regulations, financing streams, and other requirements that affordable housing providers face makes it imperative for all of these stakeholders to work together to create an effective and sustainable solution.

Solar activists have also worked to connect residents directly with programs to help them go solar. In 2015, DC's Department of Energy and Environment and the DC Sustainable Energy Utility (DCSEU) jointly announced a new incentive program that would enable low-income homeowners to receive free solar installations. The Solar Advantage Plus Program was allocated $1.4 million, with the goal of completing 130 to 140 installations. CPN held public meetings to help educate the community about the incentive program and help families apply for funds through the program. CPN also worked to ensure that the DC government made the program as transparent and accountable to community members as possible.

Fighting for the Electricity System of the Future

The emergence of solar energy as a practical choice for individuals comes at a time of significant change in the electricity sector. Across the country, the industry and regulators are in discussions about the "grid of the future." Essentially, they are trying to rewrite the rules around how electricity is generated and distributed. Who owns it? Who profits from it, and how? Who controls decision making? How are costs and benefits allocated? The critical challenge for energy activists is to identify critical pivot points and to seize these opportunities.

America's electricity generation and transmission infrastructure, and the policies that govern it, are designed for large, centralized generation plants. In this model, the financial benefits of electricity production are also centralized: revenues come from the ratepayers, and profits go exclusively to the utilities. We have recently entered an extraordinary period in which it is possible to fundamentally change this system. By shifting to a distributed energy system that incorporates rooftop and community renewables, energy efficiency, energy storage, and demand-side management, we can more widely distribute the benefits of our electricity system, decarbonize a huge sector of our economy, and lower costs for all ratepayers. We are on the tipping point of a profound transformation, but it will be an uphill battle. The status quo is heavily invested in preventing change.

Decisions being made by regulators today and over the next decade will determine our energy future. Unfortunately, this type of regulatory action is generally quite obtuse. The issues decided upon are usually too complicated to even get coverage in the newspaper, much less engage the public in an active discussion around who benefits from those decisions. The experience of Community Power Network (CPN) is that the best way to transform the system is to start by encouraging more people to become involved in a concrete and tangible way. Only then do the arcane rules around solar renewable energy credits, portfolio standards, interconnection, net metering, virtual net metering, microgrids, and community solar turn from gobbledygook into the keys to a clean energy future. CPN has found that once people start taking ownership of their corner of the clean energy economy, even if just in a small way, like putting solar on their home or school, they become true and active stakeholders in the extensive rule making unfolding all around them. A small act of one individual installing rooftop solar becomes a powerful first step in building an energy democracy constituency (figure 11-4).

Distributed energy resources, primarily solar, give electricity consumers the possibility of being producers as well. This incentivizes solar supporters to participate in discussions about the electricity system of the future, and to ensure their voices are heard. The Maryland Public Service Commission opened a regulatory proceeding in the summer of 2016 to consider the future of the state's electricity grid. Solar advocates are working to make this process as inclusive as possible, to ensure that Maryland's future electricity system works for and benefits everyone. The commission has received public comment and is now considering how the eighteen-month proceeding will be structured.

The DC Public Service Commission opened a proceeding on grid modernization in 2015. The city's solar advocates are fighting for an electricity system that is clean, local, equitable, affordable, and reliable (CLEAR). The DC PSC often gives a superficial nod to "stakeholder engagement" and goes on to implement plans designed by the energy industry. Solar advocates are working to directly involve stakeholders from all walks of life, particularly low-income residents, in this process so they can voice their own priorities, concerns, and aspirations for the District's future electricity system.

Figure 11-4. DC resident John Nugent shows off solar system installed on his roof in 2015.
Image source: Community Power Network

The fight over the electricity system of the future is taking place in many places and in many forms. One fight emblematic of the struggle between the old and new electricity system took place recently in Washington, DC, when the utility giant Exelon attempted to take over the local utility, Pepco (the Potomac Electric Power Company). Exelon is an energy behemoth. It owns electric distribution services in the Philadelphia, Baltimore, and Chicago markets. But that is only half the story. It also has significant holdings in energy generation, largely in nuclear power. By comparison, Pepco is strictly a distribution utility, meaning it does not own any electricity generating plants.

Exelon offered to buy Pepco in the spring of 2014. The utility giant hoped to secure a captive customer base (DC and Maryland ratepayers) to help make up for its financially floundering nuclear electricity generation business. As a regulated monopoly, Pepco can remain financially healthy by raising its electricity rates to cover its costs. Exelon wanted Pepco's guaranteed rate of return to offset the company's losses in its nuclear generation fleet.

This acquisition spelled big trouble for the region's solar market. Exelon has a long track record of opposing locally generated clean energy, as it sees distributed solar as a competitive threat to its electricity generation business. These two factors compelled the District's solar supporters to fight the acquisition. After many years of organizing projects and policy advocacy, DC SUN (Solar United Neighborhoods) had a strong and diverse base to bring to the campaign and quickly became the backbone of a new citywide coalition formed to fight the merger: PowerDC.

PowerDC based its opposition to the merger on the idea that rates, reliability, and integration of renewable energy are inextricably linked. The coalition brought together ratepayer advocates, tenant advocates, climate activists, and local leaders from across the city to work side by side with solar advocates in the fight over our future electricity system.

PowerDC quickly organized an outreach campaign to turn citizens out to testify at a series of public hearings in December 2014 and January 2015. Our broad, deep, and motivated coalition surprised the utilities, catching them flat footed. They attempted to drive a wedge between low-income advocates and clean energy advocates, and were surprised to learn that both sides were united. Clean energy and equity were inextricably intertwined. PowerDC turned out hundreds of activists to hearings and events and generated thousands of letters to the DC Public Service Commission. In contrast, Pepco and Exelon had virtually zero public support at the public hearings.

Volunteers went to every Advisory Neighborhood Commission (ANC) meeting in the city, visiting many of them three or even four times. ANCs are neighborhood-elected boards that represent the interests of the local community. At each meeting, PowerDC members discussed the threat of the merger and the need for citizens to get involved. Neighbors and commissioners were outraged. Twenty-seven of the forty ANCs passed resolutions opposing the merger. Pepco and Exelon began to send their executives out into the community to combat the PowerDC activists, but they failed to get any ANCs to pass resolutions in support of the merger. In an extremely divided city with contentious local politics, the public was remarkably united in opposition to the merger. After the

PSC's first review of the case, commissioners rejected the merger. It is likely that PowerDC's ANC outreach strategy was critical this outcome, as the PSC charter directs it to strongly consider ANC views.

PowerDC's persistent organizing turned the merger from a foregone conclusion into a real fight. This fight served as strong validation of the projects-to-power-to-policy model developed by DC's solar co-ops. Each co-op has community spokespeople and a network of activists from which to draw. The co-ops felt responsible to represent their constituents, to articulate their viewpoint, and to turn out their community. This enabled solar supporters to turn out a nearly inexhaustible supply of volunteers and community leaders ready to fight for solar. In the end, the strength of our coalition, framing, and strategy forced Pepco and the DC government to adopt our three goals—rates, reliability, and renewable energy—as the goals of the merger and the energy system at large.

What had once been seen as mutually exclusive goals, clean energy and lower rates, were now seen to be inextricably interconnected. Although PowerDC ultimately lost the fight to prevent the merger, the struggle became synonymous with a broadly shared alternative vision of the electricity system of the future. In this vision, local deployment of rooftop solar is optimized to create local jobs, lower grid infrastructure costs, and increase resilience to storms and other dangers.

Conclusion

Investing time and energy in community-based projects has helped more people gain access to solar, while building political power for the clean energy movement. In places like West Virginia and Virginia, solar supporters have reframed the solar debate. Solar co-ops have moved the conversation from "Does solar even work?" to "How can I get it?" even in the most conservative rural counties. In Washington, DC, and Maryland, solar advocates have changed the conversation to focus on access and equity by pushing for community solar and low-income solar programs. It is not enough that the solar market be big: it must be fair and its benefits must be widely distributed.

Educating and empowering local community members to fight for a voice in their energy future not only helps to create a better system for everyone, it creates a diverse new leadership that will continue to fight for energy democracy well into the future. Moreover, a truly distributed, democratic, and equitable clean energy movement will be unstoppable.

1. A renewable portfolio standard (RPS) is a state-level piece of legislation that sets required percentages of renewable energy to be delivered by the utilities. A "solar carve-out" means that the RPS includes a specifically delineated target for solar energy. Often the RPS targets for solar have higher incentives than other types of renewable energy because solar is more expensive to develop. These solar carve-outs have been critical in the early development of rooftop solar, as the incentive is available to everybody.

2. A solar renewable energy credit (SREC) certifies that a MWh (megawatt hour) of renewable energy was produced by an electricity generating facility. Renewable energy producers can unbundle the SRECs from the electricity they certify and sell them separately on the market. Utilities then buy these SRECs to greenwash their fossil fuel power (to credit it as renewable) to meet renewable portfolio standard (RPS) requirements. If the market price of SRECs is high enough, it encourages or subsidizes electricity generators to produce more renewable energy. However, the SREC market is very volatile, undermining the value of SRECs in encouraging long-term investment in renewable energy production. For more information, see Al Weinrub and Dan Pinkel, *What the Heck Is a REC?*, July 2013, http://www.localcleanenergy.org/what-the-heck-is-a-rec.

3. University Park Solar LLC, *University Park Solar*, accessed January 11, 2017, www.universityparksolar.com.

4. "Community Solar in Maryland," *MD SUN*, accessed January 11, 2017, http://www.mdsun .org/community-solar-updates.

CHAPTER 12

Conclusion: Building an Energy Democracy Movement

DENISE FAIRCHILD

The chapters of this volume argue that to *actually* save the planet—and ourselves—we need to get beyond the scientific and technological solutions for going to 100% renewable. Detoxing the atmosphere and oceans is only one part of the solution to climate change. Indeed, we must transform the cultural, economic, and political conditions at the heart of the climate crisis.

The contributors to this volume offer the conceptual, policy, organizing, and implementation models to begin a transformation from a fossil fuel economy to one that is sustainable, economically just, and inclusive. Collectively, the authors present a new environmental, economic, and social paradigm for energy and start to define different pathways for implementing it.

The starting point is the adoption of a new ideological frame and view of the relationship between energy and the environment. Several authors advance traditional, indigenous beliefs that our energy and other natural resources are assets that should serve a public purpose as opposed to private gain. They recognize energy—fossil fuels and renewables—not as a commodity to be bought and sold, but as part of the commons—a precious global resource that must be conserved and equitably shared. At a deeper level is to acknowledge our symbiotic relationship with nature. Nature's capacity to regenerate ensures the capacity of the human race to survive.

This new environmental paradigm necessitates a new economic agenda. The current model of extraction, processing, distribution, and utilization of resources by the private sector is part of a larger "unlimited growth" paradigm. Energy assets—both fossil fuel and renewable—are extracted to fuel mass production, mass consumption, and mass accumulation of wealth by an elite class. This contrasts with a cooperative energy service model advanced by energy democracy proponents. Movement Generation, for example, articulated its framework for a "just transition" by arguing that if *economy* means "managing home," then our energy assets must serve the basic needs of the community rather than being wasted on luxury commodities or wealth accumulation.

These ideas are being tested out across the country. Organizations like the National Association for the Advancement of Colored People, Trade Unions for Energy Democracy, and the Local Clean Energy Alliance are examples of civil rights organizations, labor groups, and energy advocates, respectively, working to dismantle the fossil fuel economy and put energy assets into the hands of communities and municipal governments. In a similar vein, Co-op Power and the Community Power Network are among a growing number of organizations committed to building alternative, decentralized energy systems: built, owned, and controlled by local communities.

These energy democracy experiments are not top-down or government-sponsored efforts. They are rooted in very practical local concerns. Health issues (high rates of cancer, asthma, and other conditions); escalating energy costs; the lack of transparency and accountability in the utility sector; and the growing impacts of climate change are moving people to act. They are spurring marginalized communities especially to be the most passionate climate and clean-energy advocates and innovators. The contributions to this volume illustrate that advocates are organizing and educating communities on the intersection of energy, climate, health, and the economy. They are shaping public policy to bring energy investments to long-neglected climate-impacted neighborhoods ("green zones"). They are leveraging the assets of anchor institutions, such as hospitals and even public housing agencies, toward improving community health, economic sustainability, and climate resilience.

Clearly, transforming and dismantling our global energy and economic system are David and Goliath–scale battles. The struggles reflected in this volume have had to overcome the financial and political power of an entrenched fossil fuel economy. These efforts taken together, however, build the foundation for a movement that is needed to beat the climate clock and radically transform the fossil fuel economy. This is particularly salient in light of current U.S. government efforts to retreat from prior environmental and climate commitments and advances in favor of fossil fuels as the energy of choice.[1]

If we are serious about climate change, we need to dismantle the fossil fuel economy and replace it with a moral economy that values ecosystems, sufficiency, justice, and real democracy. And that kind of transformation will not come without struggle. History offers a model for this kind of transformative change: the dismantling of the slave economy in the nineteenth century. Understanding the centuries-long abolitionist movement offers insight into the vision, the structural changes, the personal commitments, the political struggles, and the global movement required to stave off catastrophic climate change.

Parallels Between Transforming the Slave and Fossil Fuel Economies

The abolitionist movement offers a playbook for advocates working for climate, economic, and social justice. That movement challenged the very foundation of the global slave economy by dismantling the pillars that supported it: **property rights, profits, and power and privilege.**

PROPERTY RIGHTS

The abolitionists successfully challenged the idea that some people were property to be bought, sold, and owned. Building a sustainable and just economy requires a similar shift in thinking about nature.

The driver of climate change is an extractive economy rooted in exploiting and commercializing the environment. The Earth's natural resources—water, minerals, forests, the atmosphere—are enslaved to the global market economy in a way that is analogous to Africans under the slave economy. Like human slaves, our natural resources are devalued and chained to private interests by legal structures.

The right to extract Earth's natural resources—fossil fuels, water, timber—for private gain is fiercely fought and frequently protected through the courts and public policies and in public opinion. Oil, gas, and mining interests are waging a vigorous "Free the Land" movement through its nonprofit organization Federalism in Action (FIA) and its American Lands Council (ALC) Foundation.[2] The goal is to lease public lands for cheap[3] and to remove federal land protections by transferring federal lands to the states.[4] The Utah State Legislature, for example, is one of several western states using the courts to sue for state ownership (and privatization) of 32 million acres of federal lands in Utah. At risk is 90% of 245 million acres of federal lands currently unprotected from the extraction industry.

The water wars of the Environmental Protection Agency (EPA) are also emblematic of the entrenched and immutable commercial value and political venom accompanying efforts to protect our natural resources. Advocates for water are losing the battle against private property rights in the U.S. courts. In 2016, for example, twenty-seven states took legal action against the EPA's latest effort to define and protect the Waters of the United States (WOTUS) against private property rights. Opponents of the EPA ruling charge that it is "unconstitutional," "communism" and a "land grab."

The abolitionists faced a similar challenge. Dismantling the slave economy required a long, global struggle to outlaw the right to own, control, and exploit African labor for commercial gain. Whether or not the U.S. Constitution directly sanctioned and defined slaves as property is debated. What is clear, however, is that three clauses in the Constitution clearly permitted exploiting African slaves for their commercial value: the three-fifths compromise (Article I, Section 2); the slave trade clause (Article I, Section 9), and the fugitive slave law (Article IV, Section 2). But those "rights" fell to a constitutional challenge, and ultimately to the Thirteenth Amendment, which outlaws the right to own slaves.

Similarly, dismantling the fossil fuel economy requires challenging the right to own, extract, and exploit the environment as personal property. These rights are scattered throughout the Constitution, with private property protections supported by "due process," the "takings" clause, and "contracts," found in the Fifth and Fourteenth Amendments and in Article I of the Constitution's main text. The constitutional issue of states' rights will also make it harder to defend nature against private interests if states are successful in transferring federal lands to state ownership.

A constitutional challenge and an amendment to the U.S. Constitution are essential for protecting our environment. A credible climate change movement must integrate with the efforts of the global south and the Global Alliance for the Rights of Nature (GARN), which argues that "there is no justice as long as nature is property in law."[5] This movement is a worldwide effort to challenge constitutional rights to hold nature as property and to acknowledge "that nature and all its life forms has the right to exist, persist, maintain and regenerate its vital cycles."[6] The alliance's eco-centered approach balances the needs of humans and other species without exploiting one to the detriment of the other.

PROFIT

Private profit is a fundamental, but hidden, driver of climate change. Massive accumulation and maldistribution of wealth in the slave and fossil fuel economies occur from exploiting and controlling the engines (sources of energy) that drive production. Three hundred years of free slave labor fueled the growth of the agricultural and domestic economies, only to be replaced by fossil fuels as the fuel of choice in the industrial economy.

In the antebellum South, slaves—and wealth—were concentrated in the hands of an estimated three thousand large plantation owners, creating considerable political and economic power where "cotton was king." Many northern industrialists supported the abolition of slavery in order to shift political power and wealth from the South to the emerging class of industrial "robber barons." For those industrialists, "coal was king" for fueling factories, trains, ships, and more.

Dismantling the slave economy—while partly religious and humanitarian in intent—was, in the main, a fierce struggle for power and control over the means of production and the wealth it generated. There is a lesson here for climate change advocates: as we transition our economy once again to a new source/form of energy, we must be mindful of the core economic issues.

This is likely to be a long-term struggle. Notwithstanding the moral, environmental, and other costs of fossil fuels, they have made a small group of people very rich and powerful. A study by the Institute for Policy Studies found that CEOs of the thirty largest fossil fuel companies in the United States averaged $14.7 million in total 2014 compensation, over 9% more than the S&P 500 CEO average. The highest paid was U.S. Secretary of State Rex Tillerson, former CEO of Exxon-Mobil, who earned $33 million a year in salary and stock bonuses, but not equity pay from stock buybacks. These firms' management teams have taken home $6 billion over the past five years and receive an array of incentives and bonuses for expanding fossil fuel production and carbon reserves. [7] The compensation structure is not only four hundred times greater than that of the average worker but also provides no incentive to end climate change. In fact, it encourages extraction by any means necessary, with no consideration for its environmental, human, and economic costs.

The fight for a sustainable future is therefore also a fight for economic justice. Who will own and control the assets of the renewable energy economy—the harvesting of the sun, wind, and other renewable energy resources. Issues of economic justice get lost when climate discourse is limited to incremental advances like living buildings or greening the economy.

The structural changes in the transition to a clean energy economy could be as profound as those that accompanied the transitions from the agricultural to the industrial and digital economy, including our values, lifestyles, and basic institutions. We need to widen the lens and take a holistic view of what's at stake. A growing number of climate justice advocates have framed these changes as a "just transition," seeking to create a sustainable economy that is fair and inclusive for everyone. A just transition shifts energy monopolies to "energy democracy": community-owned and controlled renewable energy that is treated as a public "commons."

POWER AND PRIVILEGE

Finally, the transition to a sustainable future requires grappling with questions of power and privilege—who has it, how it is used, how it is distributed and controlled. Energy democracy entails such social justice concerns.

The slave economy created a society of haves and have-nots separated by race, class, gender, and privilege. The U.S. Constitution, for example, counted African slaves as three-fifths of a person. Notwithstanding the larger premise that all men are created equal, the slave economy baked structural inequalities into all aspects of society. The Constitution, laws, and informal sanctions denied African Americans access to citizenship, voting rights, education, health, family life, quality housing, food, clothing, language, religion, culture, and more. These denials were essential to maintaining power and control over property and profits.

Dismantling the slave economy was the earliest effort to eradicate such privilege and inequities. The ratification of the Fourteenth Amendment to the Constitution, in 1868, granted citizenship to "all persons born or naturalized in the United States." Unfortunately the vestiges of inequality persisted postslavery and have adapted to support the power and privilege of the fossil fuel economy. Dismantling the fossil fuel economy must entail another effort to contest all the ways that our institutions support inequalities. Again, there are parallels between slavery and the fossil fuel economy:

- Religious institutions once ordained dominion over slaves as Divine Providence; similar doctrines sanction human dominion over nature.
- Pseudoscience is used to justify privilege: just as slaves were deemed inhuman and intellectually inferior, "scientific experts" now claim human-induced climate change is a hoax.
- Educational institutions institutionalize power and privilege through

textbooks that transfer culturally biased "knowledge and values" in favor of privileged classes.

- Laws and legal institutions are used to protect property rights and discriminatory practices that serve the affluent.
- Financing institutions are used to build and maintain grow power and privilege through preferential lending.

Building a Transformative Movement

If the abolitionist movement teaches us anything about how to save ourselves from climate change, it is this: we need a movement for transformative societal change. It won't be easy. In some ways, we are all slaves to the fossil fuel economy. It is imbedded in all aspects of our economy and lives and entails a deeply entrenched culture and mindset. "Abolition" from that economy requires changing norms, values, and strongly held beliefs about property, profit, power, and privilege.

Energy democracy activists are like the abolitionists: they are building a growing awareness, advocacy, and practice that anchor a new movement with new values about property, profits, power, and privilege. But to scale this work we must identify the other prerequisites that propelled the success of the abolitionist movement. A transformative movement must:

- **Be global.** Like the slave economy, the fossil fuel economy is global and requires a global movement to change. Frederick Douglass took his abolitionist movement to Europe to gain moral authority, to influence public opinion, to raise money, and to set precedent—Europe was the first in the Western Hemisphere to abolish the slave trade. The fact that the fossil fuel economy is driven by oil producers around the world similarly suggests that this is a global struggle. In addition to Western partners, the global south particularly represents a new set of allies in the struggle, as evidenced by its growing strength and influence at the 2015 United Nations Climate Change Conference (COP 21).
- **Be multicultural.** It took the joint efforts of blacks and whites to abolish slavery. While low-income communities of color are most vulnerable to climate impacts, the fact that no one is immune to climate

change suggests the power and plausibility of a multiracial, multiethnic, multiclass, and intergenerational coalition.

- **Be multidimensional.** No one organization, no one sector of the economy, no one initiative will turn the tide. It took all the muscle—social, political, legal, and financial—of the domestic and international community to abolish slavery, as it will to build a just net-zero energy economy.

- **Be a big tent.** Cross-issue organizing is critical. We must find common ground and build alliances across movements to address the larger changes in the social and economic environment, including engaging with advocates for worker rights; Black Lives Matter; the Dreamers and other immigrant rights groups; the Occupy movement; criminal justice reform; and other proponents of all forms of social, environmental, and economic justice. As long as we stay divided, we will not build the megamovement needed to be truly transformative.

- **Offer a new meta-narrative.** The energy democracy movement must lead with a powerful moral center, offering a narrative that brings together all segments of society. Can we find a moral center that meets people where they are, that makes the case that we will either sink or swim together? The abolitionists talked about freedom and liberty as the undeniable moral authority. And for those not persuaded by altruism, many, including Frederick Douglass, used practical hooks such as the localism arguments of Thomas Jefferson, which suggested that slaves would never be good stewards of the land unless they owned and cultivated it without duress.

   What is the narrative of energy democracy? This must be a narrative that not only changes consciousness but also changes the economic paradigm—where the human economy is sustained by a healthy global ecosystem.

- **Mount legal challenges.** The fight for energy democracy will also require challenging (1) aspects of our Constitution that undermine indigenous ideas of "the commons" in favor of private property, and (2) global trade agreements that seek to replace local sovereignty with the sovereignty of multinational corporations._

- **Organize for resistance.** Wherever we are, like the cadre of global abolitionists that forced the issue of slavery into national debate, we must speak up for change. Most important, the abolitionists were strategic in their intent and methods. They used the newspaper—the media of the time—to good effect. They organized local and global networks to increase awareness, buy-in, and money. They enlisted the church community for divine intervention and galvanized speakers, including former slaves, to graphically relay the cruelty of slave life. And through the Underground Railroad and other support networks, and through protest and rebellion, the abolitionist movement built opposition to the slave economy.

- **Build alternatives.** Above all else, while resistance provides opposition to the status quo, offering alternative models of what a post-carbon and just world would look like is crucial. This is where, in addition to framing a holistic vision, we offer models of resource development, organizational structures, and institutions that convey possibilities for zero energy and water, green jobs with labor standards, healthier environments, full inclusion, and democratic practice in decision making for the new sustainable future.

The challenges to building an energy democracy movement are great. But, we don't have an option to sit on the sidelines. We are facing an existential crisis on many levels. The energy democracy movement is the unfinished business of all prior movements for justice and equality and holds promise for not only stemming climate change but advancing a global society worth living in.

1. The recent appointment of former ExxonMobil CEO Rex Tillerson as the U.S. secretary of state is one signal of the continued and, perhaps, heightened U.S. domestic and foreign policy agenda dedicated to the extractive economy.
2. American Lands Council Foundation, *Public Lands*, http://www.federalisminaction.com /wp-content/uploads/TPL-Booklet.pdf.
3. In a June 16, 2015, *Washington Post* article, Jayni Foley Hein noted that oil companies are drilling on public lands for the price of a cup of coffee—$1.50 an acre/year, https://www.washingtonpost.com/posteverything/wp/2015/06/16/oil-companies-are

-drilling-on-public-land-for-the-price-of-a-cup-of-coffee-heres-why-that-should-change
/?utm_term=.373e5bf314c7.
4. Research by the Property and Environment Research Center (PERC) and direct-action
groups—Federalism in Action and the American Land Council (ALC) Foundation—
supported by oil interests are advancing federal land transfers to help states realize
financial benefits of the extractive industry.
5. Global Alliance for the Rights of Nature, http://therightsofnature.org/what-is-rights-of-nature/
6. Ibid.
7. Sarah Anderson, Sam Pizzigati, and Chuck Collins, *Executive Excess, 2015: Money to Burn*
(Institute for Policy Studies, September 2, 2015), http://www.ips-dc.org/executive-excess-2015.

Contributors

Isaac Baker is a cofounder of Resonant Energy, a low-income solar social enterprise that spun off from Co-op Power in 2016. As a young social entrepreneur working to build clean energy resources for all, he has experience in renewable energy finance, community capital, and cooperative business development. Isaac began at Co-op Power in 2009, supporting the launch of Northeast Biodiesel and the Co-op's energy efficiency programs. Now, as co-president of Resonant Energy, Isaac works to develop community organizing and financing strategies to bring solar to low- and moderate-income households across Massachusetts and New York.

Lynn Benander is founding president and CEO of Co-op Power, a decentralized network of consumer-owned energy cooperatives in New England and New York. She is a community entrepreneur supporting the development of enterprises that work for economic, environmental, and social justice. Previously, as CEO of the Cooperative Development Institute, she supported development of cooperatives and other group-based businesses in agriculture, food, energy, and housing. "Climate change is bringing people together," says Lynn, "to find a new way of living together on this planet, in a way that is sustainable, just, and deeply democratic."

Strela Cervas is codirector of the California Environmental Justice Alliance (CEJA) and has led its growth into California's most powerful environmental justice coalition. She manages CEJA's Energy Equity program, helping communities across California chart their own vision of a clean energy future, and has led the passage of legislation in California that is unique in advancing energy policies with a focus on environmental justice communities. Before coming to CEJA in 2008, Strela worked for eight years with the Pilipino Workers Center (PWC) in Los Angeles, organizing low-wage Pilipino caregivers. She currently sits on PWC's board of directors and, along with women leaders from other organizations, helped launch the first California Domestic Worker Bill of Rights campaign. Strela is an alumna of the Women's Policy Institute.

Ben Delman is communications manager for Community Power Network (CPN). He is an experienced solar communications professional who has worked to generate attention for product and project announcements for companies throughout the solar value chain. At CPN he manages internal and external communications for more than a dozen active solar co-ops and for the organization's solar advocacy work. Along with solar advocates across the county, Ben has helped educate the public about policies that allow solar energy to compete on a level playing field with other energy sources. Ben holds a BA degree in political science from The George Washington University.

Denise Fairchild, PhD, is the inaugural president and CEO of the Emerald Cities Collaborative, where she advances its "high-road" mission to green cities that build economically just economies and ensure equity in regional economies. Denise has a forty-year track record in sustainable and community economic development, domestically and internationally. Most of her life has focused on building capacity and opportunity for South Los Angeles and other low-income residents in LA. This mission has included her training programs at Los Angeles Trade-Technical College (LATTC); her work at the affiliated nonprofit community development research

and technical assistance organization Community Development Technologies (CDTech); her directorship of the Los Angeles office of the Local Initiatives Support Corporation (LISC); and her activism in numerous civic
organizations.

Anthony Giancatarino is a fellow at the Movement Strategy Innovation Center. He provides organizing and strategy support to community-led energy
and just transition efforts in Philadelphia and supports efforts to develop
strategy and alignment regarding the centrality of racial and economic justice within the national energy transition movement. Previously, Anthony
spent seven years at the Center for Social Inclusion (CSI), working on the
Food Equity and Energy Democracy Programs, and published a series of
research reports and case studies elevating the work of communities of color
to become decision makers in the renewable energy economy. Anthony has
a BA degree in theology and political science from the University of Scran-
ton and an MPA degree from New York University.

Diego Angarita Horowitz is a data-driven professional working in innovative markets to counter climate change. He served as Co-op Power's
outreach and membership services manager for two years and has also
served in leadership roles on Co-op Power's board of directors, represented Co-op Power on Northeast Biodiesel's board of directors, and
chaired Energia's board of directors. Earlier, he worked as the associate
executive director for Nuestras Raíces in Holyoke, Massachusetts, where
he led food policy organizing efforts. Diego holds an MBA degree from
Kenan-Flagler Business School at the University of North Carolina in
Chapel Hill and a BA degree from Hampshire College.

Vivian Yi Huang is the campaign and organizing director for the Asian
Pacific Environmental Network (APEN), where she organizes immigrant community members to develop collective power and leadership
for social justice. As a child of Chinese immigrants, she is dedicated to
fighting for immigrant justice and power. Prior to working at APEN, Vivian advocated for policy, legislation, and budget funding for immigrants,

communities of color, and migrant farmworkers at Asian Americans for Civil Rights and Equality, the California Primary Care Association, and the Presidential Management Fellows Program. Vivian has also been a mentor for the Women's Policy Institute, a trainer at the School of Unity and Liberation (SOUL), and a lecturer at San Francisco State University's Department of Health Education.

Derrick Johnson serves as executive director of One Voice and vice chairman of the National Association for the Advancement of Colored People (NAACP) national board of directors. Currently he serves on the boards of the Mary Reynolds Babcock Foundation, the Center for Social Inclusion (CSI), and the Congressional Black Caucus Institute. One Voice, Inc., is a nonprofit civic engagement organization based in Mississippi and is focused on building and supporting the active participation of disenfranchised communities across the South. In his capacity as executive director of One Voice, Derrick cofounded the Mississippi Black Leadership Institute (MBLI) with Congressman Bennie G. Thompson. MBLI serves as a training program for community leaders and public officials twenty-five to forty-five years old to ensure a continuous pipeline of progressive leadership across the state. Derrick is also the founder of the Electric Cooperative Leadership Institute (ECLI), which provides education, support, and training to co-op member-owners, empowering them to exercise their ownership stake and better direct their economic futures.

Ashura Lewis is the communications and content writer for One Voice, a nonprofit civic engagement organization based in Mississippi. A native of Jackson, Mississippi, Ashura taught high school English, drama, and creative writing in both Louisiana and Mississippi public school districts for nearly a decade. At One Voice, she is responsible for print and digital content generation, communication campaign strategies, social media, branding, and other event/activity communication efforts. Ashura earned her bachelor of science degree in psychology from Jackson State University and her juris doctor degree from Mississippi College School of Law. She is a member of the Mississippi Bar Association and a published novelist.

Cecilia Martinez is director of research programs for the Center for Earth, Energy and Democracy (CEED), where her research focuses on the development of energy and environmental strategies that promote equitable and sustainable policies. She has previously held positions at Metropolitan State University, Minnesota; the American Indian Policy Center; and the University of Delaware's College of Earth, Ocean, and Environment. She has worked with organizations ranging from local grassroots groups to international organizations and has led a variety of projects addressing local and international sustainable development. Cecilia received her BA degree from Stanford University and her PhD degree from the University of Delaware's College of Urban Affairs and Public Policy.

Michelle Mascarenhas-Swan is a staff and collective member of Movement Generation's Justice & Ecology Project, which inspires and engages in transformative action toward the liberation and restoration of land, labor, and culture. Movement Generation is rooted in vibrant social movements led by low-income communities and communities of color committed to a just transition away from profit and pollution and toward healthy, resilient, and life-affirming local economies. Michelle helped launch the Climate Justice Alliance and the Our Power Campaign. As director of the Center for Food and Justice in Los Angeles, she led the launch of one of the nation's first farmers' market salad bars and the National Farm to School Program. Michelle is a former Kellogg Food and Society Policy Fellow.

Anya Schoolman is the founder and executive director of the Community Power Network, a national nonprofit that helps communities join together, go solar, and fight for their rights. In 2007, Anya founded the Mt. Pleasant Solar Cooperative with her son Walter and then fostered the creation of eleven other solar co-ops in Washington, DC, eventually forming a citywide organization, DC Solar United Neighborhoods (DC SUN) to make solar accessible and affordable to all city residents. In 2009, Anya was honored as the Maryland, DC, and Virginia Solar Energy Industries Association (MDV-SEIA) Solar Champion, as well as one of CALFinder's

10 Amazing Activists in the Name of Solar. In April 2014, she was selected as one of 10 White House Champions of Change for Solar Deployment for her groundbreaking work to deploy low-income and community solar in the National Capital Region.

Sean Sweeney is director of the International Program for Labor, Climate and Environment at the Murphy Institute, City University of New York. He coordinates Trade Unions for Energy Democracy, a global network of forty-two unions from sixteen countries. Since 1987, Sean has been involved in college-level trade union and worker education at Hofstra University; as director of the Queens College Worker Education Extension Center; and as Cornell University's director of labor studies in the Extension Division of the School of Industrial and Labor Relations in New York City. Working with unions, including the Steelworkers, Sean organized the first major conference on unions and climate change in May 2007. He also helped the International Transport Workers' Federation develop its path-breaking perspective on climate policies for transport workers. Sean holds a PhD degree in sociology and industrial relations from the University of Bath, England.

Maggie Tishman, local director of Emerald Cities New York, is committed to creating economic opportunity and wealth in low-income communities while also helping those communities address climate change. Her work with the Emerald Cities Collaborative is part of the Bronx Cooperative Development Initiative, a broader initiative to build wealth and ownership among low-income residents and people of color in the Bronx. Maggie holds a bachelor's degree in urban studies from the University of Pennsylvania and a master's degree in city planning from the Massachusetts Institute of Technology (MIT). She is currently a postgraduate fellow at the MIT Community Innovators Lab (MIT CoLab), where she is exploring how community energy projects can create economic savings and high-road jobs for local residents, as well as create wealth and enhanced energy security through community ownership.

Al Weinrub is coordinator of the Local Clean Energy Alliance (LCEA), the San Francisco Bay Area's largest clean energy coalition. LCEA sees the development of local renewable energy resources as key to growing sustainable business, advancing social equity, and promoting community resilience. He is the author of the highly acclaimed report *Community Power: Decentralized Renewable Energy in California*, has conducted energy policy briefings for many organizations and national conferences, and has led several nationally based campaigns related to science, labor, and social justice. Al is coordinator of the statewide California Alliance for Community Energy, serves on the Steering Committee of the Oakland Climate Action Coalition, and is a member of the Sierra Club California Energy-Climate Committee. He is a former national officer and a current member of the National Writers Union, UAW Local 1981, AFL-CIO.

Miya Yoshitani is executive director of the Asian Pacific Environmental Network (APEN), which she joined in the mid-1990s as a youth organizer and then worked in organizing, development, and strategic planning. APEN develops the leadership and power of low-income Asian American immigrant and refugee communities and brings their voices to the forefront of environmental health and social justice fights in the San Francisco Bay Area. APEN has won numerous policy solutions for the community and has challenged multinational corporations to mitigate the pollution that devastates the health and well-being of countless low-income communities of color. Miya participated in the First National People of Color Environmental Leadership Summit in 1991 and helped draft the Principles of Environmental Justice, a defining document for the environmental justice movement.

Index